TENDERING AND ESTIMATING PROCEDURES

John A. Milne
FRICS, FSVA, AIQS, DipEd

GEORGE GODWIN LIMITED
The book publishing subsidiary of
The Builder Group

First published in Great Britain 1968
under the title *Builders' Estimating Simply Explained*
By George Godwin Limited

Second edition 1971
Third edition 1980

© John A. Milne 1980

George Godwin Limited
The book publishing subsidiary
of The Builder Group
1–3 Pemberton Row
Fleet Street
London EC4

British Library Cataloguing in Publication Data

Milne, John Andrew
    Tendering and estimating procedures.—3rd ed.
    1. Building—Estimates
    I. Title II. Builders' estimating simply
    explained
    692'.5   TH435

ISBN 0-7114-5583-X

While the information given in this book is the product of careful research,
and is believed to be accurate at the date of going to press, neither the
author nor the publisher can accept any legal responsibility or liability for
any errors or omissions which may be made.

Made and printed by The Garden City Press Limited
Letchworth, Hertfordshire SG6 1JS

CONTENTS

FIGURES

TABLES

PREFACE

An appreciation of the factors of production and their effect on building costs is an essential requirement at all levels of management within building firms and for professionals such as quantity surveyors, architects and engineers who work in association with the construction industry. Good estimating practice helps to ensure that costs generated in the production of a building will be adequately allowed for in the compilation of price which will form the basis of the tender submitted as an offer to carry out the work. In today's highly competitive industry, with ever increasing costs for material, labour and plant, efficiency in the organisation and management of work on site and in the estimating and control of costs are prerequisites of a healthy firm.

The main aim in the previous editions of this book was to explain in a logical sequence the component parts of an estimate and to give a clear and methodical approach to pricing. This aim is perpetuated in this edition. By studying worked examples from the various trades, together with the chapters on pre-tendering procedures, profit and oncost, materials, productivity studies and labour costs, mechanical equipment and preliminaries a good basis for the understanding and building up of unit rates for preparing estimates of cost should be acquired.

The opportunity has been taken in the present edition, however, to enlarge the scope of the book so as to give an appreciation of the building industry, methods for the selection of contractors and tendering arrangements. The title has consequently been changed to reflect this greater width of content. In addition, the text has been enlarged both in descriptive elements and in the number of worked examples given and the content layout has been rationalised so as to improve the flow of information. The introduction of the sixth edition of the Standard Method of Measurement has necessitated a revision of the tables of labour outputs as well as the wording and pricing of items of work. A new chapter —Woodwork—replaces the previous chapters on Carpentry and Joinery.

The text of this book has been designed to meet the requirements of the syllabus of the Technical Education Council course in Tendering and Estimating and also to form a sound basis of knowledge for students studying for examinations in degree and diploma courses in building and quantity surveying. The majority of professional institutions who set their own examinations for membership require a knowledge of estimating. The more important of these bodies are the Royal Institution of Chartered Surveyors (Quantity Surveyors Section), the Institute of Building and the Institute of Quantity Surveyors.

The chapter on Cost Control has been primarily included to act as a guide to estimators and principals in the smaller firms who wish to improve their business efficiency, while the other techniques described are equally pertinent to them as they are to the students of the industry.

The labour outputs used by contracting firms should be built up from experience rather than obtained from books and, therefore, only the more common items of work which are likely to be required by students have been provided.

It is recognised that within a comparatively short space of time the current rapid rate of inflation will outstrip the level of cost for labour, materials and plant which have been used throughout this book. This fact is accepted but the aim, which is to lay down principles, will not be invalidated by changing levels of costs and prices.

I wish to express my thanks to friends and colleagues who have assisted me in the preparation of this book, in particular Harold Wight ARICS, James P. Morris FRICS, ACIArb and Alex Forbes MIOB, and also to the British Standards Institution and the National Joint Council for the Building Industry for permission to reproduce extracts from their publications.

John A. Milne

CHAPTER I

INTRODUCTION

In order to have an overall understanding of builders' estimating it is essential to have at least an outline knowledge of the construction industry and contract procedure.

THE CONSTRUCTION INDUSTRY

(a) Manpower

The construction industry in Great Britain employs over 1 650 000 people directly as operatives on building sites or in an administrative, professional, technical or clerical capacity within construction firms. The industry is also indirectly responsible for the employment in the manufacture, assembly and distribution of materials and components and for independent professional and associated staffs such as quantity surveyors, engineers (civil, structural and environmental) and architects in both private practice and public sector departments (eg local authorities, Department of the Environment, nationalised industries).

Table 1 shows the distribution of the workforce between private contractors and public authorities.

The distribution of the workforce employed by private contractors and carrying out works in both the public and private sectors as well as repair and maintenance work is shown in Table 2. It can be seen that just over 30 per cent of all operatives are employed on repair and maintenance work.

(b) Structure

There are approximately 80 000 separate firms which range in size from the small jobbing firm to the large national and international organisations. From Table 3 it can be seen that a very large number of small firms employ under one-fifth of the total workforce, while a comparatively small number of

1

Table 1 Construction manpower [1]

Thousands

	Employees in employment					Self-employed		All manpower
	Contractors		Public authorities			Working proprietors [3]	All	
	Operatives	APTC	Operatives	APTC	All			
1974 [4]	874	240	233	110	1456	117	427	1883
1975	820	240	233	113	1407	105	375	1782
1976	788	237	236	114	1374	93	341	1715
1977	756	229	230	112	1326	86	323	1649
1978	750	231	224	111	1315	89	368	1683

[1] Estimates by the Department of the Environment based on quarterly returns from contractors (including British Steel Corporation) and half-yearly returns of public sector direct labour. These estimates differ from those published by the Department of Employment, which do not include construction employees in building and civil engineering establishments run by authorities whose major activity is classified to some other industry (eg national and local government, etc).

[2] APTC are administrative, professional, technical and clerical staff.

[3] Figures of 'working proprietors' are for the self-employed in firms on the statistical register: figures of 'all self-employed' include estimates of those not so covered. 'All self-employed' estimates linked to results of 1975 Labour Force Survey, conducted by Department of Employment to be revised when 1977 results available.

[4] Figures from July 1974 are based on a more comprehensive register of firms: earlier figures adjusted to take account of this discontinuity. Source: DOE Housing and Construction Statistics.

Table 2   Operatives employed by contractors [1]: by type of work

Thousands

| | New housing | | Other new work | | | All new work | Repair and maintenance | | | | All repair and maintenance | All work |
| | | | | Private | | | Housing | Other work | | | | |
	Public	Private	Public	Industrial	Commercial			Public	Private			
1975	115	84	185	96	110	590	115	56	60		231	821
1976	125	86	173	87	94	565	108	55	60		223	788
1977	118	82	155	84	87	526	112	55	63		230	756
1978	105	88	143	88	89	513	114	55	68		237	750

[1] Figures include construction employees of the British Steel Corporation. *Source*: DOE Housing and Construction Statistics.

Table 3   Number of firms and distribution of workforce

Number of operatives	0–13	14–34	35–114	115–299	300–599	600–1199	1200+
Number of firms	65 450	9000	4000	1000	300	160	90
Percentage of workforce	19	14	18	12	9	8	20

very large firms employ one-fifth of the workforce. Approximately 50 per cent of the workforce is employed in firms with 115 or more operatives.

Around one-third of the workforce is employed in small firms, which produce less than 5 per cent of the total annual value of new building work but are responsible for a large proportion of maintenance and small modernisation works. The large contractors, on the other hand, who employ over one-third of the workforce, produce over 50 per cent of the total annual value of new building works but possibly under 10 per cent of the value of maintenance work.

In recent years the contractors in the middle range of size have been most at risk due to the increasing complexity and scale of building projects and the cost and availability of resources (ie finance, skilled management, labour and materials). With the reduction in building activity which has been experienced in recent years (Table 4 shows the value of output at 1975 prices) contractors from both ends of the spectrum, in order to survive, have been forced to compete for the type and size of contracts which are normally associated with medium sized contractors. This has led to bankruptcies and amalgamations; the latter allows a rationalisation of resources which helps place the new firm in a stronger position to tender for large scale projects. There has also been an increase in the number of self-employed labour only firms.

(c) Output

The total value of building output in 1978 was £16 093 000 000·00, of which over 50 per cent was public investment. The largest single area of new building output was in public and private housing. Approximately one third of all building output is in the repair and maintenance of existing buildings. Table 5 gives the breakdown of building output at current prices.

Public expenditure is either by:

(i) Central government departments such as the Department of the Environment and the Department of Agriculture.

5

Table 4 Value of output [1]

At 1975 prices seasonally adjusted *£ millions*

	New housing		Other new work			All new work	Repair and maintenance			All repair and maintenance	All work
				Private			Housing	Other work			
	Public	Private	Public	Industrial	Commercial			Public	Private		
1974	1345	1711	2714	1183	1424	8376	1896	1252	613	3761	12 137
1975	1482	1543	2511	1174	1291	8001	1658	1228	531	3417	11 418
1976	1639	1645	2489	1120	1137	8029	1530	1162	528	3220	11 249
1977	1489	1555	2371	1293	1135	7843	1594	1173	574	3341	11 184
1978	1397	1760	2291	1382	1264	8095	1860	1316	710	3887	11 981

[1] Output by contractors, including estimates of unrecorded output by small firms and self-employed workers, and output by those public sector direct labour departments classified to construction in the 1968 Standard Industrial Classification. *Source:* DOE Housing and Construction Statistics.

Table 5 Value of output [1]

At current prices

£ millions

| | New housing | | Other new work | | | All new work | Repair and maintenance | | | All repair and maintenance | All work |
	Public	Private	Public	Private Industrial	Commercial		Housing	Other work Public	Other work Private		
1974	1122	1467	2140	1073	1243	7 046	1504	992	491	2987	10 032
1975	1482	1543	2511	1174	1291	8 001	1658	1228	531	3417	11 418
1976	1795	1798	2748	1212	1226	8 780	1788	1358	618	3765	12 544
1977	1751	1867	2783	1545	1340	9 285	2106	1547	755	4409	13 694
1978	1777	2410	2953	1820	1635	10 596	2632	1878	988	5498	16 093

[1] Output by contractors, including estimates of unrecorded output by small firms and self-employed workers, and output by those public sector direct labour departments classified to construction in the 1968 Standard Industrial Classification. *Source*: DOE Housing and Construction Statistics.

(ii) Local authorities (regional and district) for housing, roads, schools, clinics, etc.

(iii) Regional health boards for hospitals.

(iv) Public corporations such as the electricity, gas and coal boards, British Rail and British Airways.

Private expenditure is for private use or investment in housing, industrial or commercial developments.

(d) Firm organisation

There are five main types of firm:

(i) General building contractors who carry out works in all the major trades.

(ii) Civil engineering contractors who specialise on roads, drainage systems and similar works.

(iii) Trade contractors who concentrate on the work of one trade, such as plumbers, heating and ventilating engineers, plasterers, bituminous felt roofers, etc. Such firms frequently work as sub-contractors to the main contractor.

(iv) Specialist contractors who specialise in a particular type of work as against a trade, eg steel or concrete frame erectors, lift installations, demolition work. Many specialist contractors manufacture the components as well as carrying out the necessary works for erection or installation.

(v) Labour only sub-contractors. There has been an increasing tendency in recent years for competent craftsmen to commence business on a labour only basis. Such craftsmen or firms will not enter into a general building contract with a client or accept responsibility for the materials which are used. The advantage of such arrangements is that they allow the main contractor to have the use of skilled labour at known costs. There is the disadvantage in that the use of a labour only workforce for part of a contract may adversely affect the organisation and management of the site.

(e) The building team
The normal parties concerned with the construction of a new

building are the client, architect, quantity surveyor and the building contractor. Each will be considered in more detail.

The client   He is the most important person in the operation as it is he or his company who are financing the project. He must feel sure that the building is designed to fulfil his requirements and will be built at an economical price.

The architect He is the person who designs the building. This is usually done in collaboration with a team of specialists, ie quantity surveyor, structural, heating and electrical engineers. The architect must study the client's needs in order to produce a building which satisfies the functional requirements and is also aesthetically pleasing. The design is developed and working drawings are prepared. The working drawings are passed to the quantity surveyor for the production of a bill of quantities and later to the contractor to enable him to build to the architect's design.

The quantity He is responsible for ensuring that the
surveyor   architect receives realistic cost advice throughout the design stage. During the working drawings stage he commences preparing the bill of quantities. The bill of quantities, together with the letter of tender, is sent to a selected number of contractors for pricing in order to receive a competitive price for the works. The quantity surveyor reports to the client and/or the architect on the tender prices and on costs generally throughout the construction of the works on the site.

The building He is responsible for the erection of the
contractor   building in accordance with the architect's drawings. His tender would normally be the lowest received offering to execute the works. The tender figure would have been arrived at

by the detailed pricing of the bill of quantities. The preparation of the tender figure requires considerable work to be done by the contractor's estimating department.

The contract documents

The documents which form the basis of most building contracts are the drawings of the proposed building, the conditions of contract applicable to the contract, the preambles and bill of quantities, a letter in the form of a formal tender from the contractor to the employer offering to carry out the work in accordance with the foregoing documents and the letter of acceptance from the employer to the builder.

The drawings Copies of the contract drawings must be available for the building contractor to inspect if he so desires, prior to him submitting his offer. It is better, however, if copies of small-scale plans and elevations are sent to the contractor together with the bill of quantities when he is invited to tender. Copies of the contract drawings are signed by the parties to the contract to certify that they are the drawings relevant to the contract.

The conditions of contract These are usually a standard form of contract conditions, the most usual being the JCT Standard Form of Building Contract (private and local authority editions) and the GC/Works/1, the form used by central government departments. The parties to the contract should sign the building contract (in Scotland) or the articles of agreement (in England and Wales). Where the new 1980 main form is used, the new JCT sub-contract form should be used for nominated work. For non-nominated work, the NFBTE/FASS 'blue' form will continue to be used (though this will require amendment to bring it into 'line with the 1980 main form).

10

The preambles and bill of quantities	The preambles give the type and standard of quality of materials and workmanship required to be used in the execution of the contract. The bill of quantities gives the description of each item of material and labour required to execute the work together with the quantity involved. The estimator is able to price each item in order to calculate a figure for carrying out the contract.

The following is an extract from a bill of quantities, the prices in the right-hand column being worked out by the estimator:

BILL OF QUANTITIES

Brickwork and blockwork   £   p

			£
Half brick walls in cement lime mortar with 10 mm beds and joints	m²	250	£7.89
One brick wall do	m²	130	£15.78
Extra over common brickwork for facing brickwork, key pointed as the work proceeds	m²	80	£1.80
100 mm Concrete block partition build in cement lime mortar	m²	75	£6.52
Hessian base bituminous sheeting damp proof course bedded in cement mortar on brick walls, overlapped 75 mm at all joinings	m²	25	£2.27
Hole for large pipe through one brick wall	No.	8	£2.50

Letter of offer	This is in the form of a formal tender which is prepared by the quantity surveyor and sent to the contractor along with the bill of quantities. The contractor must use this tender form when submitting his offer.

The following is an example of a tender form:

TENDER

Address................................

....

Date................................

Messrs. A, B and C,
Architects.

Sirs,

I/We hereby offer to execute the Several Works of the proposed offices and flats at Greenacre Road, Newtown, all in accordance with drawings prepared by you and under your direction and superintendence and on the terms and to the extent of the Bill of Quantities signed by me/us as relative hereto and subject to the Conditions of Contract included in the relative Bill of Quantities and to your satisfaction for the sum of

....

....

I/We understand that the said works shall be completed and ready for use within .... months from the date on which I/We receive intimation from the Architects that the work can be proceeded with.

I/We understand that the Employers do not bind themselves to accept the lowest or any tender and I/We declare that this offer will be held open for acceptance for twenty-eight days from the date hereto.

Your acceptance of this offer will be binding on me/us.

Dated this .... day of 19

Name .... Signature ....

Address .... Witness ....

.... Witness ....

INTRODUCTION

Letter of
acceptance
This is a letter from the client, or from the architect on his behalf, to the contractor accepting his offer to execute the works. A formal contract may be signed if either of the parties so desires.

ARCHITECTS

To Contractor.   Date ....
Dear Sirs,

Proposed Offices and Flats at
Greenacre Road, Newtown

We are authorised to accept and do hereby accept your tender dated 5th October 1979, offering to execute the Several Works of the above for the sum of X Thousand, Y Hundred and Z Pounds (£   ).

The Work will be commenced immediately on receipt of instructions to do so and will be carried out in accordance with drawings prepared by us and under our direction and superintendence and on the terms of the relative bill of quantities and will be subject to the conditions of contract included in the relative bill.

Kindly acknowledge receipt of this acceptance.

Yours faithfully,

CHAPTER II

METHODS OF TENDERING

CONTRACTOR SELECTION

In most building contracts the contractor is selected on the basis of competitive tendering. Although it is widely agreed that competition is important, the method by which it is obtained is the subject of considerable debate, the usual means being open or selective tendering. Both the Simon Committee (Report on the Placing and Management of Building Contracts, 1944) and the Banwell Committee (Report on the Placing and Management of Contracts for Building and Civil Engineering Work, 1964) criticised the undue use of the open tendering method. When the use of normal competition is not appropriate then direct negotiation may be carried out with a single contractor, or where some form of competition is introduced to arrive at a level of prices, then up to three contractors may be involved initially.

(a) Open tendering

In this method the building client advertises in the local, national or technical press, giving brief details and key information of the proposed works and inviting interested contractors to apply for the relevant contract documents. In order to reduce the number of enquiries it is frequently a condition that a small sum of money be deposited until the receipt of a bona fide tender, when it will be returned. Inviting contractors to tender for work in this fashion does not bind the client to accept the lowest or any offer (such a clause is usually included in the contract documents). The advertisement is an invitation to offer by the client; the completed document forms the contractor's offer and a contract is only completed if the client accepts the contractor's offer.

Although this method is mainly used by local authorities (although not to a large extent or to the exclusion of other

14

methods) it has been criticised in a number of government reports and is no longer used by central government departments. It has the *advantages* of:

(i) Allowing any interested contractors to tender. It therefore gives the opportunity for an unknown contractor to compete for the work.

(ii) Allowing the tender list to be made up without bias. This is the aspect which attracts local authorities who, because of public accountability, wish to demonstrate that they obtained the best bargain possible for public money and have shown no favouritism in selecting contractors.

(iii) Ensuring good competition. Since there is no obligation to accept an offer it follows that offers received are from contractors interested in carrying out the work.

(iv) Preventing contractors forming rings (ie agreeing on the level of prices to be submitted and a share out of the work so that prices are kept high) in that a lower offer could be submitted by a contractor outside the ring.

The *disadvantages* are:

(i) That tender lists can be long, thus involving a large number of contractors in pricing where only one can be successful. This excessive cost of tendering must then be passed on when pricing other works, as it will form part of the contractor's overhead charges. It is also wasteful of a skilled estimator's time.

(ii) Public accountability may be questioned if the lowest offer is not accepted. It is sometimes difficult to accept the lowest tender when the reputation of the contractor is not known, and because of the limited time available in which to make enquiries the wrong decision may be made. It is easier not to accept the lowest tender if the contractor has proved unsatisfactory on other works. This could mean that the successful contractor is ill-equipped in management, financial and/or material resources to carry out the work.

(iii) It is difficult for the client not to accept what he

considers to be a bargain. Yet too low prices affect the whole industry. The contractor who prices too low will inevitably go bankrupt and this will cause delay and prove costly to the client. In addition, such contractors cause havoc within the industry by helping to depress the prices of the efficient contractors to unnaturally low levels.

(iv) If the lowest contractor has a poor management structure then the contract can drag out and cause delay in the completion of the works.

(v) If the price is too low and the contractor realises that he is losing money then he may try to reduce the quality of the work or submit numerous claims in order to recover part of the loss. In either circumstances this will lead to a deterioration of relationships.

(vi) Many of the better contractors will not price on an open tendering basis unless they are forced to, due to lack of work. This can mean that only the less able contractors will be competing for the work.

(b) Selective tendering

In this method a short list of contractors is drawn up and they are invited to submit tenders. The number of contractors on the list should comply with the recommendations of the Code of Procedure for Single Stage Selective Tendering as published by the National Joint Consultative Committee for Building (ie the maximum number of tenderers for contracts up to £50 000 is five; £50 000 to £250 000 is six; £250 000 to £1 million is eight; £1 million plus is six). The short-list can be drawn up by the client's professional advisers or alternatively an advert, giving the details of the proposed works, could be placed in the press requesting contractors to make application to be considered for the work. A short-list of the most suitable contractors is drawn up and the remainder informed of their non-inclusion on the list. Two contractors, however, should be kept on reserve in case any contractor in the short-list should drop out prior to tenders being submitted. Many local authorities maintain lists of contractors who are willing to undertake work of a specific type, within certain cost limits, and in particular geographic localities. Such lists may be

prepared through advertising in the press or from knowledge
of contractors who have undertaken work in the past (Figure

REGIONAL COUNCIL

SELECTIVE
TENDERING

Applications are invited from firms wishing to be included on a list of
Contractors from which selections to tender on a restricted basis will
be made. The list will be divided into the following categories:

1. BUILDINGS

(a) Several Works—Building
(b) Demolition Works
(c) Site Investigation
(d) Playing Fields and
Landscape Work
(e) Painting and Decorating
(f) Heating and Ventilating
(g) Electrical

2. ROADS

(a) Civil Engineering—Road
and Bridge Works
(b) Site Investigation—Road
and Bridge Works
(c) Masonry Walling
(d) Kerbing and Concrete Slab
Paving
(e) Fencing
(f) Thermoplastic Carriageway
Markings

3. WATER

(a) Civil Engineering—Service
Reservoirs, etc.
(b) Pipeline Construction
(c) Site Investigation

4. DRAINAGE

(a) Civil Engineering— Sewage
Treatment Works, etc.
(b) Sewer Construction
(c) Site Investigation

5. PLANNING

(a) Rehabilitation of Derelict
Land

6. SUPPLIES

(a) Printing Work

The list will be graded according to size of contract, as follows:

GRADE A – Contracts over   £750,000
GRADE B – Contracts between   £350,000 and £750,000
GRADE C – Contracts between   £75,000 and £350,000
GRADE D – Contracts between   £25,000 and £75,000
GRADE E – Contracts up to   £25,000

**Application Forms are available from the Director of Law and
Administration.**

The approved list will be kept under continuous review and firms
may apply at any time for inclusion.

Figure 1

1). These lists should be kept under review. This method has the *advantages* that:

(i) As only competent contractors were invited to tender then the lowest tender can be accepted.

(ii) It reduces the risk of failure and cuts the cost of preparing estimates.

(iii) It enables competing contractors to include an adequate level of profit which in turn helps to give stability to the industry.

It has the *disadvantages* that:

(i) Care requires to be taken to ensure that favouritism does not influence the inclusion or exclusion of firms from particular lists.

(ii) There is a reluctance to amend lists (ie it can be difficult to get onto an established list and/or there is a reluctance to remove unsatisfactory contractors from a list due to the consequential ill-effect this would have on their trading position).

(iii) Tender prices are invariably higher than they would have been under open tendering. This increase should be a reflection of better management, ensuring that completion dates are met and a higher standard of workmanship achieved.

(iv) Contractors who receive documents may submit high prices rather than return the documents unpriced in case their name is removed from subsequent tender lists.

(v) There is a greater chance of collusion between firms unless the composition of the list is changed for each contract.

(c) Serial tendering

This is a useful method of contract when the client has a continuing building programme. Contractors are invited to tender in competition on the basis of a normal bill of quantities, but on the understanding that a series of contracts for similar work will be let to the successful contractor on the conditions contained in that bill of quantities. This type of contract is particularly useful when there is a similarity of building types, such as in housing, as the contractor gains familiarity with the organising and management while the workforce become more efficient due to repetition. This method has the *advantages* that:

(i) It allows the client and the contractor to programme their workload in advance with more certainty.

(ii) It generally leads to better relationships between the contractor and the client and/or architect and lends itself to a situation where the contractor may offer advice in the planning of future works (ie to maximise the utilisation of formwork or plant; advice on standardisation or prefabrication of units so as to reduce the time taken for the works).

(iii) It allows the contractor more time to plan the work on the site so that it can be carried out more efficiently.

The *disadvantages* are that:

(i) It relies on the integrity of the client in that, since he knows the contractor's prices, the follow-on contracts may be based, to an unfair extent, on the low-cost items in the bill of quantities.

(ii) It reduces the work available, under competition, to other contractors who may have wished to price.

(d) Negotiated tenders
Negotiated contracts are usually entered into for particular reasons, eg the contractor has special management skills or can undertake particular works which require a high degree of technical competence, or is capable of completing the works within the required, restricted time period. If there is no special reason for a negotiated contract then some other form of contract is likely to be more suitable for the client.

Under a normal negotiated contract using a bill of quantities the contractor is selected at an early stage in the design process. This early selection, however, need not exclude competition, as this can be achieved by means of a 'two-stage' procedure. Preliminary competition can be on the basis of a tender document in which the contractor indicates the level of costs required for labour rates, site establishment charges, general overheads and profit.

After acceptance and as the drawings are prepared, the contractor and the quantity surveyor will prepare and price a

bill of quantities for the work. The amount of detail in the bill of quantities will be agreed by both parties and the level of pricing will be that laid down in the initial tender document.
The *advantages* are that the contractor:

   (i) Can give advice to the architect during the development of the design.

 (ii) Can commence ordering materials, prefabricating work and programming so that an early start can be made on site and production can flow smoothly.

The *disadvantage* is that the cost of the work is likely to be higher than for a competitive tender.

TENDERING ARRANGEMENTS

There are a number of forms of contractual relationship which may exist between the client and the contractor but they tend to fall within two broad categories:

 (1) Cost reimbursement contracts where there is no risk to the contractor and little incentive to ensure that the work is carried out in an efficient manner and at an economically sound price.

 (2) Price given in advance contracts where the price is agreed before the work commences and the contractor carries the risk. There is therefore an incentive for the contractor to ensure that the work is carried out in an efficient manner.

Within each category there is a wide variety of contractual methods. For all methods there are both advantages and disadvantages but each has a valid use and is suitable in the correct context for the right situation.

COST REIMBURSEMENT CONTRACTS

General
Provided that a contractor of good repute is employed then contracts on this basis are suitable for work:

 (a) Which is required to be carried out at short notice such as may be necessary in the case of urgent repair and/or maintenance work.

 (b) Which due to the uncertainty as to the extent and true

nature of the work, such as may arise in alterations and repairs to existing properties, the exact content cannot be predetermined.

(c) Where the client wishes to exercise a measure of control over the method of working or to supervise the work such as may be necessary in the co-ordination of machinery installation where builders and sundry works are required.

(d) Where the contractor possesses a special ability which the client/architect wishes to utilise, such as highly skilled tradesmen for intricate work or those of exceptional quality.

(a) Cost plus percentage
In this type of contract the contractor is paid the actual prime costs of the work plus an agreed percentage. Care must be taken to define what costs are to be regarded as prime costs, as disagreement can arise on whether certain items should be included in the percentage addition (eg is travelling time to be paid as a prime cost or included in the percentage addition). The percentage addition should include for all extra costs incurred by the contractor including overheads and profit. The definition of prime cost, as published by the Royal Institution of Chartered Surveyors and the National Federation of Building Trades Employers, is a useful basis for this type of contract, but its use is neither mandatory nor implied.

One percentage may be applied to the total of the prime costs or separate percentages may be used for labour, materials and plant.

The *advantages* are:

(i) That the basis of the contract can be agreed and the work commenced straightaway, thus avoiding the delay which is necessary if estimates have to be prepared.

(ii) That if the contractor is efficient then the cost to the client should represent a fair price for the work undertaken. The client knows the percentage addition which is being applied to the prime costs.

The *disadvantages* are:

(i) That there is no incentive for the contractor to be efficient in his use of labour, materials or plant as he will be paid for these costs plus the agreed percentage. Indeed, an inefficient contractor may gain more from this type of contract than a better organised firm.

(ii) It is difficult to predict at an early stage of the works the final cost.

(iii) Some form of control/supervision is necessary in order to ensure that the correct labour and plant hours are being charged for by the contractor. Such control may be difficult and/or costly.

(b) Cost plus fixed fee
This type of contract is similar to the previous one, but instead of a percentage being added to the prime cost of the works a fixed lump sum is added. The intention is to reduce wastefulness in that it is to the contractor's advantage to complete the work as quickly as possible since his fee remains constant. It therefore has the *advantage* in that it removes the incentive to prolong the time taken to complete the work. This, however, leads to the *disadvantages* that:

(i) The advantage of speed in commencing the work is undermined since before a fee can be agreed a fairly detailed scheme of work must be prepared.

(ii) If there are any major variations, which can be difficult to avoid in this type of work, then the fee must be re-negotiated to take account of such variations.

(c) Target cost
In this type of contract a target price is agreed for the works. If, when the works are completed, the actual cost is less than the target cost, then the difference in cost is shared between the contractor and the client on a pre-agreed basis. If, however, the actual cost is greater than the target cost, then the contractor is usually only paid the actual costs plus an agreed percentage to cover his overhead costs. The target cost is frequently negotiated and agreed from a priced bill of quantities.

This type of contract has the *advantage* that the contractor has an incentive to carry out the works as quickly and as economically as possible. The client also stands to benefit through the contractor's efficiency. The *disadvantages* are:

(i) Difficulties may arise in agreeing on a revised target cost if there are numerous' variations during the contract.
(ii) Administrative costs may be high due to the need to check the actual prime cost of the work as well as to prepare a bill of quantities.

PRICE GIVEN IN ADVANCE CONTRACTS

General

The forms of contract in this category are the most common for medium and large works. The risk to the contractor is influenced by whether the contract is on a fixed price or fluctuation price basis. Only contracts of under twelve months' duration, however, should be let on a fixed price basis (ie there is no fluctuation clause in respect of the cost of materials and labour) and the contractor must allow in his offer for any possible increases in costs during the period of the contract. Although the client has the advantage of having a known cost for the works, in periods of high inflation it may be cheaper for the client if he retains the risk by using a fluctuation contract. The market conditions prevailing at the time will influence this type of decision. In a fluctuation cost contract the contractor is reimbursed for increases in prime costs during the contract from those ruling at the date of the tender. The final cost is not accurately known until near the end of the contract due to the difficulty in assessing the likely incidence of inflation on building costs.

(a) Lump sum or plan and specification contract

The contract document in this type of contract comprises drawings, conditions of contract and specifications. The contractor is responsible for taking off the quantities and preparing his estimate. The onus is therefore on the contractor to include in his price everything necessary for carrying out the work.

For small contracts where the client's requirements have been fully established and there are good drawings and specifications, then this can be a useful form of contract. It has the *advantage* that the client knows the total cost of the work.

The *disadvantages* are:

(i) That as all competing contractors are required to take off their own quantities, it involves them in a great deal of work when only one can be successful. If sub-contractors are asked by the main contractor to quote then, if four main contractors are pricing, it is possible that as many as twelve sub-contractors may be competing for a portion of the work, and the lowest sub-contractor may not be successful if the main contractor he priced for did not submit the lowest offer.

(ii) That unless the information given is very good, and this can be difficult, then all contractors may not be pricing on an even basis. The contractor who prices for a lower or inferior specification, when the documents are not specific, or knows from past experience what the architect or client will accept, has a better opportunity of submitting the lowest offer.

(iii) If a contractor makes a mistake in his taking off from the drawings then he is responsible and will not be allowed reimbursement under the contract.

(iv) Since only a lump sum price has been submitted by the contractor (usually on a trade basis) he can often rely on official extras to make up any discrepancies between the tender price and his costs for the job. Having no detailed breakdown of costs makes the agreement of the prices charged for extras difficult for the client and/or his professional advisers.

(v) In order to cover for the extra risks involved in this type of contract, the contractor is likely to include his own contingency sum which will be under his control. The architect and/or client will be unaware of its extent and exercise no control over its expenditure.

Some of the above disadvantages can be overcome if the specification is presented in two parts:

(a) Materials and workmanship which refer to British Standard Specification, trade literature and codes of practice, and

(b) Description of work which presents the work in a series of numbered items, each of which is required to be priced. This method is a compromise between a bill of quantities, where the individual items would have less work content and have quantities attached, and a traditional specification which tends to be a wider description of the works.

(b) Contracts based on a schedule of rates
In this method a specification and a list of the more important items of work in each trade are sent to the contractor for pricing. The item descriptions and the units of measurement are similar to those used in a normal bill of quantities, but no quantity is given. This method has the *advantages* that:

(i) It gives the level of costs which will be incurred by the client and forms the basis for agreement of other items of work performed by the contractor during the course of the work.
(ii) Work can be commenced earlier than if a full bill of quantities had been prepared.

The *disadvantages* are that:
(i) The client has no indication of the final cost of the works.
(ii) It is very difficult to determine which contractor submitted the most advantageous offer unless approximate quantities are applied to the rates so that totals can be compared.
(iii) The contractor's price may have been influenced by the quantity involved (ie if the item was likely to refer to a large or small quantity of work).

(c) Contracts based on a bill of quantities
A bill of quantities comprises conditions of contract, trade preambles and measured items of materials and labour. The trade preambles indicate the quality of materials and workmanship required, while the measured items give the quantity of materials and labour necessary to complete the works.

The *advantages* of contracts based on a bill of quantities are:

(i) That it cuts down the amount of work to be done by the contractor prior to preparing his estimate.

(ii) It removes the onus of responsibility for incorrect quantities from the contractor to the quantity surveyor.

(iii) It gives a fair basis for competition in that it lays down the extent, workmanship and quality of materials on which the contractors are requested to offer.

(iv) It gives control over the amount the contractor can charge for extra works.

(v) It gives the client a good indication of the final cost of the work.

It has the *disadvantage* that it requires time for the bill of quantities to be accurately prepared.

(d) Package deal contracts

The contractor, in this type of contract, is responsible for the design as well as the construction of the building. Competing contractors are required to comply with the client's brief but are given scope to utilise their building knowledge and skills. Most contractors who undertake work of this nature rely heavily on industrial building systems or standardised prefabricated units.

This method has the *advantages* that:

(i) The design and construction skills are combined within one organisation.

(ii) The contractor, because he has complete responsibility, will give both a firm price and a completion date.

(iii) A number of different solutions can be considered by the client if several contractors are competing for the work.

The *disadvantages* are that:

(i) Within a contractor's organisation, production methods and economic considerations generally take precedence over design and aesthetic considerations.

(ii) In order to reduce the risk of abortive work, if pricing in competition, the design and specification will not be at as advanced a stage as necessary for an accurate price to be given. Contingencies which are under the contractor's control are therefore required and included in the tender figure.

26

(iii) When a number of different schemes are received it can be difficult to gauge which one represents best value for money and is the most suitable for the functional requirements of the client. To assess the different schemes and to give advice it is often necessary to appoint an independent quantity surveyor and architect.

(iv) Any variations to the design as accepted, during the course of the construction, tend to be more costly than would have been the case under more normal building contracts. This is due in part to the tight work schedules being followed by the contractor on site and the difficulty in altering or modifying an industrialised system.

CHAPTER III

PRE-TENDERING PROCEDURES

ORGANISATION WITHIN FIRMS

The organisation within building firms varies with the size and type of work undertaken. Figure 2 gives an outline of how a firm could be organised. Small firms have similar management requirements to larger firms which means that responsibility for the work is assumed by fewer staff.

Feedback from site

Estimates must be based on a sound knowledge of the firm. Labour outputs and waste allowances on materials should not emanate from price books but be based on an analysis of site costs for previous, similar work. The contracts manager, through planning and control on site, can assess the time required to carry out works and the need for overtime and/or incentive schemes. Through time sheets and returns by the bonus surveyors the average time taken to carry out the various items of work can be determined. The actual wastage of materials can be gauged by comparing the materials paid for under the contract with the invoices less any material removed off the site at completion. The plant manager can provide actual costs for the various types of plant. All this information is required by the estimator to enable him to prepare a true estimate for the cost of the proposed works. The feedback of site information is also an essential control procedure which can be used to monitor the progress of the work and check the costs. Discrepancies in either, provided that they are identified within a reasonable time, can be investigated in order to establish the cause and if possible to rectify the position. It also enables areas of inefficient working to be isolated, for instance, where outputs are significantly different from those achieved by others, and/or excessive waste of materials to be highlighted (ie due to theft

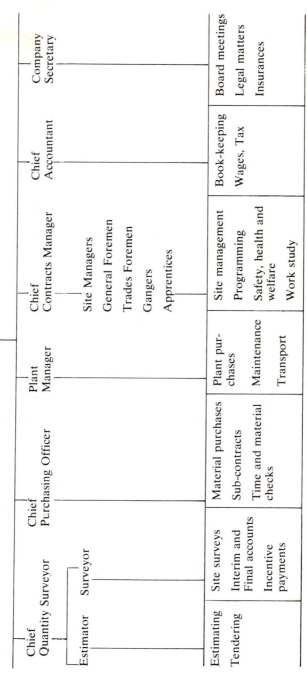

BOARD OF DIRECTORS

MANAGING DIRECTOR

Chief Quantity Surveyor	Chief Purchasing Officer	Plant Manager	Chief Contracts Manager	Chief Accountant	Company Secretary
Estimator Surveyor			Site Managers General Foremen Trades Foremen Gangers Apprentices		
Estimating Tendering	Material purchases Sub-contracts Time and material checks	Plant purchases Maintenance Transport	Site management Programming Safety, health and welfare Work study	Book-keeping Wages, Tax	Board meetings Legal matters Insurances
Site surveys Interim and Final accounts Incentive payments					

Figure 2

or short measure deliveries). Estimating and feedback are therefore an on-going process with information being constantly checked and updated.

Initial action on receipt of tendering documents

Where a contractor (or estimator) is pricing in competition, it is essential that before time is spent in the taking off of quantities for a lump sum contract or the pricing of the items in a bill of quantities, that a decision is made whether or not to submit an offer for the work. Such a decision can only be rationally made after assessing the type and extent of the work involved in relation to present workload and known future commitments. It is necessary, therefore, to study the various documents and drawings that make up the contract and to visit the site. For this the estimator requires the co-operation of others within the organisation in order to arrive at a cost for the works. The contracts manager will be required to prepare an outline programme for the work, a schedule of plant requirements and to indicate the extent of the site establishment—huts, hoardings and the like. The purchasing department will need to arrange for quotations for materials and check regarding availability and delivery dates. Works to be sub-contracted will require to be issued for pricing. The plant manager, in consultation with the contracts manager, will give estimates of plant and transport costs. The surveyor will advise on the contract documents and on particular conditions and special requirements. Some of the points to be considered may be summarised as follows:

GENERAL INFORMATION REQUIRED FOR ESTIMATING

(1) **The preliminaries**
    (a) Check that standard conditions of contract apply. Examine any amendments made to these conditions and see that they are fair and workable.
    (b) Note the starting and completion dates.
    (c) Check if there is a time limit for completion and whether there are liquidate damages for non-completion on time.
    (d) Assess when the bulk of outside work will require to be

done and consider the effect of this on guarantee time and completion dates.

(e) Check if the contract has to be executed in any particular sequence.

(f) Consider any requirements regarding special insurances in terms of the contract.

(g) Note the period of interim payments, when the final account will be certified and the limit of the retention fund.

(h) Check if the opportunity will be given to quote for any nominated sub-contractor's work applicable to his firm.

(i) Note the location of the site: if it is in the country consider transport, roads, access and availability of labour. If it is in the town consider access, unloading and storing of materials.

(j) Note the length of the defects liability period.

(k) Check if the contract is on a fixed price basis.

(2) The trade preambles

(a) Check that all the materials are available and that supplies will not interfere with completion dates.

(b) Check on any handling or storage difficulties that may be encountered if new or unusual materials are specified.

(c) Ensure that your firm has experience in handling and fixing all materials specified. If there are materials of recent innovation then investigate possible difficulties in fixing, amount of waste expected and labour requirements.

(3) The drawings and technical reports

(a) Assess the likely requirements for mechanical plant and scaffolding.

(b) Assess the relationship between the value of new work and works in alterations.

(c) Check access and working space for plant and storage space for materials.

(d) Consider the requirements for huts, welfare facilities, toilets and temporary roads.

(e) Consider the supervisory staff requirements and the approximate size of the labour force.

(f) Consider security and the necessity for hoardings and/or watchmen.

(g) Check the position of boundaries, accesses and services. Consider the number and position of temporary water points, power and/or sewerage connections.

(h) Check on the completeness of the drawings, as this could reflect on the accuracy of the bill of quantities. Also well-prepared drawings invariably imply that the scheme has been fully considered and disruption of work on the site due to numerous variations is less likely than if the scheme appears to be ill-prepared.

(i) Read all technical reports prior to visiting the site. For example an engineer's report on soil conditions may be available.

$\longleftarrow$ - - - - - $\longrightarrow$

(4) The site

(a) Check the actual position of the site in relation to adjoining property. Consider the access and the transport of plant and materials. Ensure that there are no obstacles to the transport of plant to the site, such as narrow bridges or overhead cables. Assess problems due to traffic restrictions, such as limited waiting and no parking and their effect on the unloading of materials.

(b) Determine the position of temporary roads, huts, toilets, hoardings and storage areas. Check on the suitability of the site for the use of plant.

(c) Consider the possibility of employing local labour. Check the availability of accommodation should men require lodgings in the area.

(d) Determine the type of ground to be excavated. This may be done by the inspection of trial holes. Enquire regarding the height of the water table and on the tendency to flooding. Read the relevant clauses in the bill of quantities.

(e) Note the position, distance and time taken to travel to the nearest tip and the charge for its use, if excavated or other materials are required to be removed from the site.

(f) Check the position of water, drainage, electricity and telephone services for the provision of temporary supplies.

(g) Determine the general weather conditions for the area and consider their effect on guaranteed time allowances.

(h) Check on the availability of public transport to the site, the time taken for the journey and the distance from the yard.

(i) Check, in the case of alteration work, that unoccupied property is, in fact, empty property.

The building contractor, after having looked into the implications of the contract, is now in a position to decide whether or not to submit an offer for the work.

The following points should be carefully considered:

(1) How, in relation to present and likely future commitments, will this work fit into the firm's overall building programme?

(2) Will there be sufficient workmen available to complete the work within the time limit laid down by the client?

(3) Will there be supervisory staff, who are competent to handle this work, available at the time of the contract?

(4) Do they have plant and equipment necessary to complete this contract successfully, or is it available for hire at realistic rates?

(5) Will they have the necessary financial resources available for use on this contract?

(6) Will the type of work involved make the best use of the labour force and are they competent to undertake such work?

(7) Do they require the contract in order to keep their workforce fully employed?

(8) Is there sufficient time available to prepare an accurate tender?

Insufficient men, lack of proper plant and poor management hold up contracts, frustrate the architects and lose clients. The contractor must also consider the cash flow through the firm so that at any particular time he is not in a financial situation that could be detrimental to the future of the firm.

If the contractor decides to proceed and submit a tender he must obtain quotations for the following:

(a) The cost of all materials delivered to the site.
(b) Estimates, on the same conditions as the main contract, for any portions of the work which he intends to sub-contract.

On receipt of the required quotations the estimator may proceed to build up rates for all the items in the bill of quantities or for the work requirements in a lump sum contract. He must include for the following:

Materials	The cost of materials delivered to the site. Labour unloading and storing. Waste; changes in bulk; consolidation.
Labour	The cost of the men's wages plus the cost of all the other charges and allowances that must be paid by the employer. The performance standards of the men while carrying out the various elements of the work.
Plant	The true cost of the plant to the contractor, allowing for all standing charges and running costs. Transport, erection and dismantling charges and variable costs.
Temporary works	The cost of all temporary works necessary for the completion of the building works and for removal of same on completion of the contract.
Overheads	A proportion of the cost of general supervision and oncosts necessarily incurred in running the business.
Profit	A sum of money which is commensurate with the risk and effort involved in the organising and carrying out of the works.

The information available to the estimator will contain

many uncertainties due to the fact that the work will take place some time in the future and at a site some distance from the builder's head office or yard. In order to reduce the risk of error the estimate should be built up in a methodical manner and the estimator should stick to a strict routine.

When 'all-in' quotations from sub-contractors or suppliers are received for portions of the work, such as precast concrete or specialised joinery work, then this portion of the estimating risk is removed from the main contractor, although he still has the overall responsibility to ensure that the works are satisfactorily completed. The main contractor will require to add a percentage (usually $2\frac{1}{2}$ to 5 per cent) to cover administrative costs and give an element of profit. The sub-contractor may require special facilities, such as the unloading of materials by the main contractor, storage and canteen facilities on site, the hoisting of units and the like; the cost of such work, plus profit, will require to be included in the tender figure.

General terminology
The following definitions are given as an alternative means to that described above for the component parts of an estimate.

Prime costs:
This refers to the cost of the material, labour and plant required for the works, ie the basic inputs. The prime costs of material and labour are usually allowed for in the build-up of unit rates. Plant, however, may either be included in the general overheads or priced as a lump sum in the appropriate items in the preliminaries or trade sections of the bill of quantities or in the particular unit rates in which its use is required.

Overheads:
  (a) General overheads. The term overheads generally refers to general overheads. They are the outgoings of the firm which are required for its day-to-day functioning and which are not directly associated with any particular contract, for example, head office administrative costs.
  (b) Project overheads. This refers to site costs which are incurred in order that the building may be erected in an

efficient manner but does not include the prime costs. They would include such items as site administration (ie supervision permanently on the site) and temporary works.

Special risks:
Special risks and considerations which affect a particular job will require to be isolated in order that their financial implications can be established. In contracts where a bill of quantities is provided such items will be contained in the preliminaries section.

Whereas project overheads pertain to most jobs, special risks are particular to the job under consideration.

Profit:
Gross profit. This refers to the money left after all direct costs are met. (The direct costs are the prime costs plus the project overheads.) The gross profit is therefore the proportion of general overheads charged against this particular job plus the balance of the money due.

Net profit. This is the money left after the requirement for general overheads has been met.

Tendering policy
In order to maximise their resources all firms should have a tendering policy which is based on knowledge gained from past trading. A non-selective approach to the preparation of offers to carry out work will inevitably add to the estimator's workload and impair the accuracy of his estimates. Also, tendering for and accepting work without adequate financial, managerial or manpower resources will mean that at times the value of work in hand will go beyond that which the principals of the firm can successfully control. The unnecessary additional work, even if secured at high rates, may prove to be a source of difficulty resulting in a low return or possible loss. Equally important, however, will be the effect on the profitability of other current works due to resources being spread too thinly, resulting in profit margins being depreciated so that net profits may be no greater or even less than would have been achieved with less work.

A major cause of bankruptcies in the building industry is overtrading. Turnover should not be pursued for its own sake; profitable turnover should be the aim, and this may be better achieved with reduced activity. To prosper and to remain viable it is necessary for a firm to have sound policies concerning the size and type of work they wish to consider as being their norm. This does not mean that in periods when work is scarce or the firm wishes to expand, that tenders will be submitted for other types of work or larger contracts.

Tendering policy should not be based on impressions but on an analysis of past contracts. By this means it is possible to determine the size and type of work which most frequently resulted in a profitable return. Only by having such knowledge is it possible to make the most effective use of management skills and the estimator's time. By appreciating both the strengths and weaknesses of the firm it is possible to plan more effectively and increase efficiency. Such an analysis should encompass a comparison of profitability of work undertaken for different clients, architects and, in the case of sub-contractors, main contractors.

The indecision of a client or architect may result in numerous variation orders which may disrupt the programme of work, extend the contract period and result in a drop in the level of the anticipated profit. Variations, although paid as extras, can be costly to the contractor due to their disruptive effect and because they tend to be valued in accordance with the level of rates contained within the contract documents. The frequency and level of interim payments may also have an effect on the actual net profit earned. In the case of sub-contractors, the organisation and co-ordination abilities of the main contractor can have a major influence on their profit.

Time is best spent in preparing offers for work which are most likely to achieve the greatest profit. It therefore follows that a selective approach to work undertaken is the basis of sound business procedure.

Each job should be considered on its merits, bearing in mind any adjustments which may be necessary to compensate for the general level of risk anticipated or other matters discussed above.

Geared payments

Having prepared the estimate and arrived at a tender total, it is necessary to ensure that the bill of quantities total corresponds with the tender total. In order to achieve an increased payment cash flow at the early stages of a contract, which will help to finance the work, contractors at times redistribute the money within the bill of quantities rates so that early works are priced high and later works are priced low; the tender total remaining the same. This system of forward loading the bill of quantities rates appears advantageous to the contractor but it requires extreme caution since:

(i) Any reduction in the items which are overpriced will seriously affect the profitability of the contract.

(ii) The contractor, because of receiving high interim payments, may not detect the true position regarding site costs and be lulled into a feeling of false security.

Care must therefore be taken in the choice of items which can be adjusted. For example, excavations and underbuilding heights may vary due to actual site conditions being established but concrete ground floor slabs will remain unaltered. For better control of site costs the contractor requires to use two bills of quantities—the one submitted with the tender (ie front loaded) and the one used to prepare the estimate (the true prediction of costs). Interim payments would then be prepared using both bills of quantities. The former for arriving at the amount of money due by the client for the work done to date; the latter to compare actual site costs with the amount of money actually earned for this portion of the works. Control is achieved but the amount of work required in the preparation of interim payments is increased.

The client's quantity surveyor may advise his client against accepting any bill of quantities which he considers is not a bone fide offer. This advice would be given in order to protect the client should the contractor go bankrupt during the currency of the contract. If unduly high payments have been made for the work done on site the difference between the original tender figure and the costs for another contractor to complete the contract would be far greater than normally expected, and the client would stand to lose financially.

The distinction between an estimate and a tender

A tender is an offer made by a builder to actually carry out work at a price stated. An estimate only indicates the probable price of the work without offering to do it. The estimate forms the basis on which the tender figure is derived. The object of the estimate is to predict the likely cost of the proposed works, while the object of the tender is to obtain the contract so as to earn maximum profit. The tender price that is submitted for a contract should be a management responsibility. Management must decide the amount of profit they require to carry out the contract and this may be added as a percentage or as a lump sum. Provided that the estimate is prepared in a logical manner from information on previous costs and known production outputs, it is possible to consider the proposed contract in light of the market conditions at the time of the tender and the likely workload within the firm in relation to future commitments. These factors may influence the tender figure.

CHAPTER IV

OVERHEADS AND PROFIT

In order that a business may function properly and maintain its position in the industry it must earn enough money to cover the cost of all overheads and earn a profit for the shareholders and/or directors.

This money, which is additional to the cost of labour, plant and materials required for the execution of the contract, may be added either as a lump sum or as a percentage addition.

The sum of money allowed must cover the following:

Supervision	General supervision as given by the owner, manager or directors. The cost of supervision that is permanently on the site and can be directly attributed to that site is usually charged as a lump sum against the site. The cost of a travelling supervisor who is responsible for a number of sites is usually allowed for as an overhead.
Oncosts	The costs necessarily incurred in running the business, eg office and yard rent and rates, electricity, telephones, stationery, office salaries and expenses, car expenses, advertising; interest on borrowed capital, maintenance and depreciation of office and yard buildings and equipment, accountants' fees, insurances not allowed for in labour costs, bad debts and the like.
Profit	The difference between the contract sum and that required to pay for overheads, site costs, labour, plant and materials to complete the contract.

Overheads may be divided into two categories: fixed costs

and variable costs. The fixed costs remain relatively constant, in the short term, irrespective of the level of activity in the firm. Variable costs fluctuate with increases or decreases in the level of work undertaken. The following is a list of typical overheads:

Fixed costs

    (1) (a) Office salaries —administrative staffs, including travelling foremen, secretaries, book-keepers and typists.

       (b) Indirect charges on salaries—eg National Insurance contributions.

    (2) Rent and rates. A rental charge should be made even when the property is owned. This charge would be similar to the rental income which could be obtained if a third party had occupation of the premises.

    (3) Printing and stationery.

    (4) Postages and telephone charges. This cost could increase in periods of less activity on site due to efforts in trying to attract new work and catching up with back work.

    (5) Electricity and fuel costs, ie light and heat.

    (6) Insurances. All insurances other than specific site insurances, eg employers' liability third party, fire, theft and accident.

    (7) Repairs, maintenance and depreciation.

       (a) Office building and equipment, ie furniture and machinery.

       (b) Minor plant, eg planks, trestles, barrows, shovels, hand tools, scaffolding.

    (8) Car expenses, Repairs, maintenance, depreciation and running expenses.

    (9) Professional fees, eg accountancy.

  (10) Sundry expenses, Advertising, bad debts.

Variable costs

    (1) Interest on borrowed capital, ie retentions on work in progress and overdrafts, loans etc.

    (2) Salaries of principals, ie directors' salaries which are related to profit.

Profit

Profit requirements vary from firm to firm depending, among other factors, on the size of the contracts undertaken and nature of the work (ie alterations or new work, as this could affect the length of time taken to turn over the contract value and consequently the length of time to earn the profit for completing the works). Other matters to be considered are the client, his professional advisers, the regularity and level of interim payments, and the effect this contract would have on anticipated work load.

In arriving at the amount of profit required it is also necessary to assess the level of risk inherent in the proposed form of construction and contractual arrangement.

Risk may therefore be conveniently considered under two main sections:

(a) Contractual risks. These are risks stemming from the contract documents and the necessary arrangements for work to be done by sub-contractors and/or deliveries from suppliers of materials or components. Also included would be the firm price tender risk.

(b) Technical risks. These risks centre around the form of construction (whether traditional or non-traditional) and the ease or otherwise of executing the work, previous experience of erecting buildings of similar construction, and the problems of programming and plant utilisation.

The greater the risk involved the higher the profit requirement, while the converse would apply when the risks are low (or normal).

Profit is the only item in a tender that can be safely adjusted because of policy decisions or be a subject for negotiation between the contractor and his client. Overheads are costs and as such are as real as site costs. The difference between overheads and profit should therefore be fully appreciated.

The percentage addition for overheads and profit is important and must be determined before the unit rates can be fully calculated. The following example has been designed to illustrate the method of arriving at this percentage:

EXAMPLE
The business under consideration is that of a general building

contractor who has had successful trading over a period of years. The business is under the direction of two directors and employs an average annual total of 60 operatives. Records kept over a number of years show a relationship between the cash value of wages to the cash value of materials as being approximately 40:60.

Total wages bill

		£
32 tradesmen at £70.00 per week		*2 240.00*
7 apprentices at £50.00 per week		*350.00*
21 labourers at £60.00 per week		*1 260.00*
	Per squad week	*£3 850.00*

Annual wages bill: 49 weeks × £3850.00   £188 650.00

The cost of general indirect charges on labour must also be added to cover the following:

(a) National Insurance
(b) Annual and public holiday pay
(c) Non-productive overtime
(d) Redundancy and sundry costs
(e) Construction industries training board levy
(f) Employers' liability and third party insurance

Records kept over a number of years indicate that these items amount to approximately $33\frac{1}{3}$ per cent of the gross wages bill, ie £62 820.00.

	£
Net wages	*188 650.00*
Percentage addition $33\frac{1}{3}$%	*62 820.00*
Total wages bill	*£251 470.00*

Allowing for materials in the proportion of 60:40 (materials: labour) the gross annual turnover of work would be £628 675.00

Overheads		*Per annum*
(1)	Salaries of principals *Directors' salaries including compulsory* *contributions*	£ 18 000.00
(2)	Office salaries *Office and general administrative staff at a* *combined salary of £10 000 including all* *compulsory contributions*	 10 000.00
(3)	Rent and rates *Office and yard. Rent* £1 000.00 *p.a.* *Local authority rates* £1 200.00 *p.a.* £2 200.00	 2 200.00
(4)	Office expenses *Printing, stationery, postages, telephones,* *electricity and fuel.* *Average total cost of £50.00 per week* *Cost per annum: 52 weeks at £50.00*	 2 600.00
(5)	Insurances *Allow for all insurance policies not* *included in labour costs and site insur-* *ances. These insurances would include* *fire, theft, accident, etc*	 3 000.00
(6)	Advertising *The cost of advertising*	 175.00
(7)	Maintenance and depreciation *Maintenance and depreciation of office* *and office equipment*	 400.00
(8)	Car expenses *Allowing for three cars.* *Interest charges, annual running expenses,* *repairs and depreciation*	 5 500.00
(9)	Minor plant *An allowance has been made for repair,* *maintenance and depreciation of minor* *tools, plant and equipment which is* *supplied by the contractor and the cost of* *which is not readily allocated against* *individual contracts, ie trestles, battens,*	

Carried forward 41 875.00

Brought forward *41 875.00*

tools, shovels, barrows and scaffolding for work not exceeding 3.5 m high. Capital cost of equipment—(say) £5 000.00. Allow 5% for interest, 5% for repairs and 10% for depreciation.
Cost to firm: 20% of £5 000.00   *1 000.00*

(10) Interest on capital
Under the conditions of contract (JCT) the contractor will require to lie out of retention monies until the end of the maintenance period. In addition, interim payments may not always be adequate and the final account may be delayed beyond the prescribed time limit.
The limit of the retention fund is usually 5% of the contract value.
By allowing 10% of £628 675.00 (ie £62 870.00) this covers the amount of retention monies and any necessary additional floating capital. If this money was available to the firm it could have been invested. Therefore the cost to the firm is 9% of £62 870.00.   *5 655.00*

(11) Professional fees
The cost of accountant and other fees   *500.00*

£49 030.00

The percentage of overheads to estimated annual turnover of work is

$$\frac{£49\ 030.00}{£628\ 675.00} \times 100 = 7.80\%\ approx.$$

This percentage does not allow for any of the following factors:

(a) Indirect labour charges, such as yard time, periods when on non-productive work, unemployed time and the like, when the workmen are doing work which cannot be charged against a particular contract or job.

45

This may be due to non-continuity of work for short periods through which the men are kept in employment.

(b) Reduction in workforce due to illness or scarcity of work. This will reduce the value of work done by the contractor but the cost of the firm's overheads will remain constant. Overheads charged on a percentage basis may not be fully recovered.

(c) The cost of minor remedial work done during the maintenance period. This may be allowed for under overheads or assessed and added as a lump sum in the preliminaries section of the bill of quantities for each individual contract.

(d) The amounts of bad debts that are written off.

In order to cover for expenditure which may arise from these items the overheads percentage has been taken as 9 per cent.

The percentage allowed for profit will vary and will depend on some or all of the following:

(i) The nature of the firm's business. A firm of painters and decorators will require a larger profit percentage than a firm of general builders in order to achieve a comparable profit.

(ii) The size of the contract and whether it is mainly new work or work in alterations. (See page 232.)

(iii) The area in which the work is required to be done.

(iv) The client and his professional advisers. An architect who does not prepare a complete set of drawings or issues excessive variation orders may so disrupt the contractor's programme that he requires to increase his profit margin. The same may be true if the surveyor is over meticulous or slow in finalising contracts.

(v) Whether the contractor requires the contract to give continuity of work.

In this example the profit percentage has been taken as 6 per cent.

This gives a combined overheads and profit percentage of 15 per cent. This percentage has been used throughout the book with the exception of Chapter XVIII (Electrical Installations).

Note: The following refers to examples in Chapter XVIII (Electrical Installations) only.

Using the figures from the above example the percentage addition for profit and oncosts allowed for in the chapter on Electrical Installations has been calculated as follows:

Overhead expenditure = £49 030.00
Annual labour costs = £251 470.00

$$\text{Percentage recovery} = \frac{£49\ 030.00}{£251\ 470.00} \times 100 = 19\tfrac{1}{2}\% \text{ approx.}$$

Allowing for additional items as previously described 20 per cent has been added to labour costs to cover for overheads.

A profit of 6 per cent on both materials and labour costs has also been allowed.

General

The percentage required for overheads and for profit varies between firms and between trades. The method used to recover overheads may also vary. The contractor must ensure that all costs that are not directly attributed to a particular job are in fact recovered over his year's trading—or other such period that he uses for budgetary control.

The usual methods for recovery of overheads are:

(a) Overhead expenditure = £10 000.00
     Overturn of firm = £143 000.00

$$\text{Percentage recovery} = \frac{£10\ 000.00}{£143\ 000.00} \times 100 = 7\% \text{ on job costs}$$

(b) Overhead expenditure = £10 000.00
     Annual labour costs = £65 000.00

$$\text{Percentage recovery} = \frac{£10\ 000.00}{£65\ 000.00} \times 100 = 15\tfrac{1}{2}\% \text{ on labour costs}$$

(c) Overhead charges may also be calculated using the total productive hours.

Overheads are usually calculated on the most stable factor. In electrical contracting the cost of materials can fluctuate greatly while the labour costs remain comparatively stable. Different light fittings and specialised equipment can vary greatly in cost but the labour content involved for fixing in position may be similar. Because of this, electrical contractors tend to recover overheads on their labour costs rather than on job costs. By analysing past records a contractor can determine the most appropriate method for the recovery of overheads.

Overheads and profit may be charged for as a percentage addition on each item in the bill of quantities or they may be charged for as a lump sum and inserted in the preliminaries section of the bill of quantities. The method used can have an effect on the amount of the final account. If the contract is increased in value due to variations and the profit and oncosts have been added as a percentage onto each item, then the contractor is bound to recover oncosts and profit for the additional work. If a lump sum had been charged then this may not have varied because of the increased amount of the contract. If the contract, however, had been decreased in value, then the contractor would still have recovered the lump sum charged for profit and oncosts even though the amount of work was reduced. If the percentage had been added to the unit rates then the amount of money received by the contractor for profit and oncosts would have been reduced correspondingly to the amendment of the quantities in the final measurement.

Oncosts are an important factor in the cost to the contractor for carrying out a building contract. Not only must he ensure that the amount of labour, material, plant and temporary works charged for are appropriate to his needs, but he must also ensure that the amount he receives for overheads is adequate. To do this he must budget for what his firm aims to attain over a given period of time (usually a year). The contractor must plan the anticipated expenditure and the value of work he proposes to undertake. As overheads are recovered on turnover it is essential that his assessment of turnover is realistic. The budget should consider the likelihood of expansion and whether there is enough money available. The expansion and the running of the business should be planned. A flexible budget for the firm's overheads should be prepared and used as a basis for determining the percentage to be charged.

CHAPTER V

MATERIALS

The cost of materials must include the following:

(a) The cost delivered to the site. The contractor must ensure that where materials are quoted ex works, etc, he includes for all necessary additional costs to bring the materials onto the site.

(b) The cost of unloading and storing. The contractor must allow for any double handling that may be necessary due to the nature of the site, conditions of contracts, or for his own convenience in prefabricating parts of the work off the site. Due to lack of access or congestion on a site materials may be required to be unloaded and wheeled to the site. Joinery fitments are usually made in the joiners' workshop, split into sections for transport, reassembled and erected in position on site.

Storage huts for materials are usually charged for in the preliminaries section of the bill of quantities.

(c) Waste and pilfering. Waste should be allowed on all materials and is due to the following:

(i) Breakages, eg lavatory basins get cracked, drain pipes, glass, etc, get broken.

(ii) Cuttings, eg rainwater pipes and gutters are manufactured in standard lengths and because of position of fittings, etc whole lengths may not always be used. Plywood and other sheet materials come in standard sizes that may not correspond to the dimensions required on the site, and so necessitate cutting and waste.

(iii) Depreciation due to bad storage, eg cement in bags if stored in damp conditions will go hard. Plasterboard deteriorates if damp and may become unusable.

(d) The cost of returning empty cases. If expensive cases are used to transport materials then there is usually a credit value in the case provided that it is returned.

(e) Compaction and loss of bulk. Some materials reduce in bulk when placed in position. For example:
  (i) Hardcore when compacted loses about 25 per cent in volume, and
  (ii) there is a reduction in volume between the dry ingredients for concrete and the wet mixed concrete. The percentage reduction varies with the concrete mix specified.

(f) Increase in volume. Excavated materials increase in volume on excavation and this has an effect on the cost of removal.

(g) Delivery of materials as and when required. The contractor, when buying materials, should not only be looking for the cheapest price but also for good service. If a supplier can deliver materials to the site as and when required then this helps the contractor to adhere to his programme of work and lessens the possible sources of delay. Provided that the service is good it may be cheaper for the contractor in the long run to deal with this supplier in preference to a supplier with cheaper quotations but a poor service.

Unloading materials
Many materials quoted to be delivered on site require to be unloading. The contractor therefore requires to allow in his estimate for the labour for unloading and storing these materials. This may be done by adding the cost of this labour to the unit rates or by assessing the total labour costs and allowing for these as a lump sum in the preliminaries. Materials such as sand, aggregates, and sometimes common bricks, may be tipped off the lorries and no charge made for labour unloading. Facing bricks, timber, cement in bags and the like require unloading and stacking. In the case of timber it may be easier to assess the number of loads required and the total time to unload and stack and to allow for this as a lump sum in the preliminaries. With facing bricks and cement, however, it may be more convenient to allow for this by adjusting the material quotation to include for unloading. For example:

Cost of cement d/d   £28.00 per tonne
Unload and store
0.75 h labourer at £2.70   £2.03
                             £30.03 per tonne delivered on site, unloaded and stored.

On contracts requiring large quantities of concrete, the cement is usually delivered in bulk containers and transferred to cement silos on the site. In this case no unloading charge is necessary. It is also becoming more common practice for bricks to be delivered in bundles held together by wire. These bundles may be unloaded by crane and the cost of unloading would be covered for in the lump sum cost of the crane.

The time taken to unload may be added to the labour output as shown under bar reinforcement on page 141, or may be added to the labour section of the unit rate build-up as shown in slating Example 2 in Chapter XV (Roofing).

In estimating, the method used would at all times be the one that gives the most accurate result for the type of material being considered.

Unloaded materials require to be stored on the site in positions convenient for distribution and incorporation in the building. Efficient distribution can improve productive output. Careful storage can achieve savings due to a reduction in handling costs and loss through accidental damage.

Approximate times taken to unload materials by hand:

Material	1 Labourer
Cement or plaster (in bags)	0.75 h per tonne
Bricks	2.5 h per thousand
Timber (carcassing)	1.0 h per m³
Concrete interlocking tiles	2.25 h per thousand
Copper tubing (28 to 42 mm)	1.0 h per 150 m
Drain pipes (stoneware or fireclay)	3.25 h per thousand
Reinforcing rods	3.00 h per tonne

Discounts
Material prices are generally subject to reduction due to trade and cash discounts. Some contractors, due to the amount of materials purchased through a merchant, may receive a

preferential discount. Trade discounts are generally in the order of 10 per cent to 60 per cent and cash discounts are generally 2½ per cent or 5 per cent.

The price of materials used in the build-up of material costs should be trade prices (ie the retail price less the trade discount). The cash discount is not deducted at this stage for, unless the merchants' accounts are paid within the required period (usually one month), the cash discount is not allowed. Cash discounts, however, may be considerable and if not deducted they form a hidden profit. Management often require the total of the cash discounts to be assessed so that they can ascertain the actual anticipated profit from the works. Provided that management considers it feasible to pay the accounts within the time limit and that the money from discounts is relatively safe then they may make a lump sum adjustment on the estimate so as to arrive at a tender figure. The consideration of factors such as this help management to submit tenders for contracts based on known facts in relation to their current needs.

TRADE TERMS IN COMMON USE

c & f	Cost of freight included in price.
c.i.f.	Cost, insurance and freight included in price.
COD	Cash on delivery.
D/d site	The cost of materials includes for delivering them to the site. The contractor is responsible for the unloading of materials on their arrival at the site.
E & OE	Errors and omissions excepted.
Ex works	The cost of materials is at the manufacturer's works. The contractor must pay for transport from works to the site or yard. The manufacturer loads the lorries at the works. The cost of delivering and unloading is met by the contractor.
FOB	Free on board (ship). Manufacturer pays for loading and cartage of materials to nearest port. The contractor pays for the unloading of the ship (done by others) and for transporting materials to yard or site. The contractor is also responsible for any harbour dues.
FOQ	Free on Quay. Generally as last but manufacturer pays for unloading materials and landing charges.

FOR The contractor pays for loading and transporting materials to site or yard. Free on rail. Manufacturer pays for loading and cartage of materials to the nearest railway station. The contractor unloads the rail containers and transports materials to site or yard. The contractor is also responsible for any demurrage charges (ie railway charges for not unloading within a limited set down period—usually 24 hours).

Nett cost This is the final cost after all deductions (ie trade and cash) have been deducted. It is the nett price to the builder.

Prime cost This is the cost after the trade discount has been deducted but before the cash discount is deducted. If there is no cash discount then the nett price and prime cost will be the same.

£

eg To supplying 1 tile fireplace and hearth	200.00
Trade discount 20%	40.00
Prime cost	160.00
Cash discount 5%	8.00
Nett cost	£152.00

Pro rata In proportion.

CHAPTER VI

PRODUCTIVITY STUDIES

The total quantity of labour required to perform a certain item of work is expressed as a 'labour output'. Labour outputs compiled by individual firms should be calculated from experience built up over a long period. The time taken to perform the same item of work will vary from firm to firm, gang to gang, and area to area, depending upon the type of firm and the familiarity of the gang in performing the task.

The time taken to build a square metre of brickwork, for example, should not be based on the time taken by a gang when working at full speed but must be based on the amount of brickwork built over the period of a week or on the time required to build a predetermined amount of brickwork. This would, therefore, take account of wasted time, tea breaks and the like. The labour output would not be calculated using the best or poorest gang but on the performance of an average gang.

Labour outputs or measured/quantifiable time standards are based on a unit of production with the assumption that the quality is: (a) acceptable, and (b) appropriate to the precise type and category of work.

More progressive firms calculate labour outputs by means of work study, but care must be taken in order to calculate outputs over a long enough period to ensure that they are realistic.

The quantity of materials is accurately known from the bill of quantities and provided the contractor receives a good quotation from the merchant, the cost of materials should not vary significantly from that of the other contractors competing for the work. The cost of labour, however, is gauged from past experience and it is on this portion of the cost that the contract is largely won or lost, although the end result also depends on the factors of oncost and profit required by the competing contractors. This makes the accurate determination of the amount of labour required of major importance to the contractor.

The use of work study

Work study may be used to assist the estimating process. One of the estimator's main problems is the assessment of performance standards for his workmen for the many different items of work. A complete study of these items would require the employment of a full-time work study officer and, although this would lead to a more accurate assessment of performance standards, it would also be expensive. Work study officers may not be considered a viable proposition other than in the larger contracting organisations: a small builder may, however, use work study techniques to a limited extent so as to improve the efficiency of his estimating without incurring additional overheads. The use of work study concepts of rating and standard performance make it possible to rationalise previous observations of performances into a systematic assessment of outputs. Standard performances which have been competently assessed are more likely to be accurate than information that is arrived at by the use of less systematic methods. The more accurate the information on labour outputs the less risk there is in the preparation of the estimate. Labour costs can be based on known performance standards.

Method studies may also be carried out to determine whether or not the most appropriate method is being used to perform an item of work. By carefully selecting items for method studies it may be possible to reduce the amount of labour hours required in some key items of work.

The following terms which are used in work study have been extracted from British Standard 3138: 1969 Glossary of Terms Used in Work Study and Organisation and Methods (revised 1979), and are reproduced by permission of the British Standards Institution, 2 Park Street, London W1A 2BS, from whom copies of the complete standard may be obtained.

Standard performance	The rate of output which qualified workers can achieve without over-

exertion as an average over the working day or shift provided they adhere to the specified method and provided they are motivated to apply themselves to their work. This is represented by 100 on the BS scale.

Qualified worker One who is accepted as having the necessary physical attributes, who possesses the required intelligence and education and has acquired the necessary skill and knowledge to carry out the work in hand to satisfactory standards of safety, quantity and quality.

Rating The numerical value or symbol used to denote a rate of working.

British
Standard
rating scale
A scale where 0 corresponds to no activity and 100 to standard rating. Ratings should be expressed as × BS.

Standard
rating
The rating corresponding to the average rate at which qualified workers will naturally work, provided that they adhere to the specified method and that they are motivated to apply themselves to their work. If the standard rating is consistently maintained and the appropriate relaxation is taken, a qualified worker will achieve standard performance over the working day or shift.

Basic time The time for carrying out an element of work or an operation at standard rating.

Extension The process of converting observed time to basic time, ie:

$$\frac{\text{observed time} \times \text{observed rating}}{100 \text{ (standard rating)}}$$

Relaxation
allowance
An addition to the basic time to provide a qualified worker with a general opportunity to:

(a) recover from the effort of carrying out specified work under specified conditions (fatigue allowance), (b) allow attention to personal needs, and (c) (rarely) recover from adverse environmental conditions. The amount of the allowance will depend on the nature of the work and may be taken away from the place of work under management direction.

Note. Health and Safety legislation and codes are relevant.

Contingency allowance

A measured or estimated allowance of time which may be necessary for inclusion in the standard time to cover specified and legitimate work activities and/or unavoidable interruptions (not recorded waiting time) in work sequence. Contingency allowance is applied only when it is impractical to treat such items as occasional elements.

Work content

Basic time and relaxation allowance and any other allowance for additional work (eg, work contingency allowance).

Standard time

The total time in which a task should be completed at standard performance, ie basic times plus contingency allowance plus relaxation allowance.

The rating of performance is not a scientific process but a measured judgment based on the work study officer's experience. A competent tradesman performing an item of work well in a steady, efficient manner under proper supervision would be given a 100 rating. A 100 rating is referred to as Standard Rating. Work done at Standard Rating plus appropriate allowances for relaxation, personal needs, etc, gives a performance which is referred to as Standard Performance. Few tradesmen work at Standard Performance or 100 rating unless adequately motivated to do so by an incentive scheme. A normal rating for a competent tradesman not working on bonus would be about 75 to 80. A rating of 60 could be described as very slow with the operative having little interest in his work, while a rating of 130 could be described as a fast, tradesmanlike performance of an operative working on bonus and still producing the desired standard of workmanship. It is important that the rating of operatives is carried out by a person who knows the work involved. An operative may appear to be working fast but because he is using the wrong method or tools he is not producing as much as another man who appears to be working more slowly but is using the correct method and/or the correct tools. The second man should be given the higher rating as he is working more effectively and is providing the higher output.

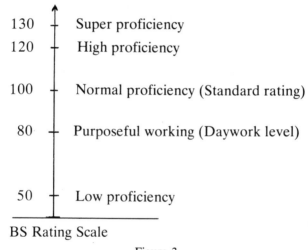

BS Rating Scale

Figure 3

Using standard ratings then Standard Times may be calculated as follows:

(a) A bricklayer laying 78 bricks per hour and rated at 120

$$\text{Standard Time} = \frac{\text{observed time} \times \text{observed rating}}{\text{standard rating}} = \frac{60 \times 120}{100}$$

= 72 minutes for 78 bricks or 65 bricks per hour

(b) A bricklayer laying 52 bricks per hour and rated at 80

$$\text{Standard Time} = \frac{60 \times 80}{100} = 48 \text{ minutes for 52 bricks or 65 bricks per hour}$$

If a building contractor is operating an incentive scheme then he could base his estimate on performance standards which corresponds to his labour working under financial motivation. This has been described as at 100 rating on the BS scale. The labour rate used for preparing the estimate would allow for the bonus earnings of the man (see section on Incentive Schemes later in this chapter). This would be the standard hour rate for the man working at 100 performance. The incentive would be a planned incentive.

$$\text{Standard Time} = \frac{\text{observed time} \times \begin{array}{c}\text{observed} \\ \text{rating}\end{array}}{\text{standard rating}} + \text{allowances}$$

The allowances are an assessment of the time the men are not actually producing. They would include for time lost due to tea breaks, going to the toilet, general conversations, correcting mistakes, waiting for materials to arrive or for other trades, etc. These items are usually referred to as relaxation, fatigue, personal needs and contingency allowances. The percentage allowance is calculated by individual firms from their records. Provided that the rating is done efficiently then the standard time should be the same, but if a bonus scheme is being operated then the amount paid as bonus to the various workmen would be different (ie their ratings are different therefore their bonus earnings are different). Instead of the men being described as slow, average or fast, a rating may be given to their performance. In this example ratings of 80, 100, 125 respectively have been given and an allowance of 25 per cent has been added to cover the various items listed above. Although the men's times varied the standard time calculated is constant.

Fast man

$$\frac{16 \text{ min} \times 125}{100} + 25\% \text{ allowances} = 20 \text{ min} + 25\% = 25 \text{ min}$$

Average man

$$\frac{20 \text{ min} \times 100}{100} + 25\% \text{ allowances} = 20 \text{ min} + 25\% = 25 \text{ min}$$

Slow man

$$\frac{25 \text{ min} \times 80}{100} + 25\% \text{ allowances} = 20 \text{ min} + 25\% = 25 \text{ min}$$

The appreciation of the value of work study and the use of standard performances as a means of determining the labour content for items of work should lead to more accurate estimating. Rating and standard performances may be used in estimating irrespective of whether or not an incentive scheme is being operated. An example of their application to estimating is given in the following section on incentive schemes.

INCENTIVE SCHEMES

The following is a statement from the Working Rules of the National Joint Council for the Building Industry on the general principles governing both the operation of incentive schemes and the making of productivity agreements:

(1) **Objects**

The objects of incentive schemes and/or productivity agreements are:

(a) to increase efficiency, thereby keeping the cost of building at an economic level, and
(b) to encourage greater productivity, thereby providing an opportunity for increasing earnings by increased effort, while maintaining a high standard of workmanship and avoiding a waste of labour and materials.

It follows that such agreements must be strictly related to productivity.

(2) **Incentive schemes—application**

The intention is that incentive schemes shall be applied generally throughout the building industry and shall cover all trades and/or occupations.

The effective application of incentive schemes depends

61

upon willing co-operation between management and operatives to ensure on the one hand that the organisation of the job is such as will permit realistic targets to be achieved and on the other hand a genuine effort is made to improve output. Where it is necessary to carry out work study this should be arranged by mutual consent.

(3) **General principles**
  (a) A target should be issued by management for each operation to be performed by an individual operative, or gang, and, according to the extent that performance is better than the target, an additional payment should be made over and above the appropriate standard rate of wages.
  (b) Targets should be issued before operations are started and, wherever it is possible to do so, they should be agreed with the accredited representatives of the operatives concerned, or with their union officer.
  (c) Targets should be based on standards of performance which have, wherever possible, been determined on jointly accepted work study principles published by the BSI.
  (d) Targets are dependent on the saving rate adopted in each scheme. The incentive scheme must state the proportion of the saving which is to be paid out as bonus.
  (e) The number of operatives to be treated as a unit for bonus purposes should be as small as is operationally practicable. Bonus should not be paid on a trade or site collective basis except where there are exceptional circumstances and it has been jointly agreed.
  (f) Incentive schemes should be expressed in simple and precise terms in order that
     (i) operatives may readily know what they have to do to increase their earnings, and
     (ii) misunderstandings and disputes may be avoided.

(4) **Operating principles**
  (a) The target should be stated as a given quantity of work to be done in a given number of hours, to the satisfaction of management. (The given number of hours may

be expressed as a monetary value where this method is customary.)

(b) Where tasks are pre-measured they should be of short duration so that, as far as is possible, they do not extend into a second payweek.

(c) Gains and losses occurring in different payweeks shall not be off-set, except where a target which has been pre-measured covers work to be done in more than one payweek.

(d) Working targets once fixed may not be altered unless there is a significant change in the job content or in working methods and then only after joint consultation.

(e) At the commencement of repetitive work a jointly-agreed 'learning-curve' allowance is permissible having regard to the improvement in productivity that should subsequently follow.

(f) The target will be inclusive for craftsmen and labourers and all hours will be chargeable against the target except where there is an interruption of work beyond the control of the parties.

(g) The time of non-working supervision should not be charged against the gang. In the case of part-time working supervision the proportion of time to be charged against the gang should be agreed in advance.

(h) The time of first-year apprentices should not be charged against the gang. In the case of apprentices in their later years of apprenticeship the proportion of their time which should be charged should, as a guide, be the same as the proportion of the craftsmen's rate which they receive under the apprentices' wage for age scalc.

(i) Overtime premiums, guaranteed time and travelling time should not be charged against targets.

(j) Bonus payments, after adjustment in the case of a proportionate scheme, should be made at the standard plain time rate of the operative concerned, including extra payments under NWR's 1.10, 1.11, 3B and 3D.

(k) The amount of bonus earnings should be notified to operatives not later than the pay-day next following the payweek in which the work was completed. The bonus should be paid not later than the next pay-day after that.

(l) Where work for which bonus has been paid proves defective and has to be re-executed in whole or in part, (i) the remedial work shall be carried out by the same operative gang, (ii) no bonus shall be paid therefor, and (iii) the time taken shall be off-set against any savings on subsequent targets. This provision shall not apply where the original work had been carried out strictly in accordance with precise instructions.

(5) **Productivity agreements**
The objective of a productivity agreement is to make a joint effort to improve efficiency by reducing unit costs through such means as the use of balanced gangs, greater flexibility or the relaxation of specified work practices. Such an agreement should provide an opportunity for high earnings.

(6) **Disputes**
   (a) In the event of a dispute or difference arising over an incentive scheme or productivity agreement, there shall be no restriction of work or withdrawal from operation of the scheme whilst the procedure outlined in this paragraph is being followed. Any settlement of such a dispute or difference shall apply with retrospective effect from the date upon which the dispute or difference was raised officially by the accredited site representative.
   (b) The dispute shall be discussed in the first place between management and site representatives of the operatives concerned in accordance with the provisions of NWR7. If these discussions are not successful there should be a meeting between management and the full-time officer of the union(s) concerned. If the dispute remains unresolved the parties, or either of them, may ask the National Joint Council to arrange for an independent investigation of and report upon the point of difficulty.
   (c) If thereafter the parties are still unable to resolve the difficulty, they shall refer it for decision to the joint industrial machinery in which event the report on the

independent investigation will be made available to the Conciliation Panel.

(d) Details of incentive schemes and/or productivity agreements should be made available, on request, to Employers and Operatives Local (or Regional) Secretaries.

DIFFERENT METHODS OF REMUNERATION AND INCENTIVES
The above principles refer mainly to premium bonus schemes, but before studying them in detail it is worth considering other methods of incentives that are used.

There are several methods in current use for calculating the earnings of employees for work done. The most common method is that based on Time-Work Rates in which the rate paid per hour is multiplied by the number of hours worked by the employee; the rates of wages and the working rules being laid down by the National Joint Council for the Building Industry. This may be a good method when quality is more important than quantity or when expensive materials are being used and where speed may have a serious effect on the amount of wastage. This method, however, has a big disadvantage in that it offers no incentive to the good worker to increase his output or to improve the efficiency of the methods used. The workmen also tend to await instruction rather than show initiative or seek further instructions from the foreman. Good supervision is, therefore, necessary in order to achieve continuity of work from employees.

The cost of labour is a major factor in the cost of building and should, therefore, be studied with a view to reducing overall costs. Low wages to employees do not necessarily mean low labour costs. Higher wages and greater efficiency may prove to be more economical since the saving in labour hours may more than compensate for the higher wages. The builder must ensure, however, that he is achieving greater efficiency when the men are earning higher wages as these themselves do not necessarily mean a greater output, other than when initially introduced. This may be done by introducing good incentive schemes.

An incentive scheme in itself does not ensure that the contractor will not make a loss but it encourages his workmen to work harder and reach the level of performance at which they can earn bonus. A good scheme will be designed so that a competent workman can earn bonus without materially affecting his standard of workmanship. The main incentive schemes in operation are:

(1) Piece Work Rates:
(2) Profit-Sharing and Co-partnership; and
(3) Premium Bonus Schemes.

(1) Piece Work Rates. Under this method the employee's earnings are related to the output he achieves. His earnings will be based on the number of units completed, multiplied by the rate per unit, irrespective of the time taken to do the work. Generally there is no guaranteed basic wage for the time spent on the job, eg:

Labour only sub-contractor being paid £10.00 per 1000 bricks laid.
Payment = 15 000 bricks laid at £10.00 per 1000 = £150.00.

This method has the advantage of increasing output and standardising the labour cost of production. Its disadvantage is that output is only increased to the extent that the workmen consider to be a reasonable level of earnings. The quality of work needs inspection to ensure that it is not reduced due to increased speed of production.

The method described is Straight Piece Work Rates, but there are variations of this such as Differential Piece Work Rates and Piece Work Rates with a guaranteed day rate.

(2) Profit-Sharing and Co-partnership. These schemes try to foster loyalty to the firm and collective effort of employees by dividing between them, at set intervals, a proportion of the profit of the business. The more prosperous the business the greater the profit. An employee's share of profit is usually related to his length of service and his annual earnings.

These schemes may be run jointly or separately. In co-partnership the profit bonus is left in the company as shares or as a high-interest loan.

The advantages of this scheme are that, provided the general wages are good then the employees will feel that they are receiving a fair deal. Morale will be good, turnover of labour low, good productivity, greater care in handling plant and equipment and less wastage of materials will be achieved.

The disadvantages are that all employees are paid profit irrespective of individual efforts, the interval between payments tends to be lengthy, ie annually or half yearly, and the interest of the workmen tends to wane, the amount of profit earned is not fully in the control of the workmen and may be influenced by good or bad management, and also a great deal must be taken on trust as all employees cannot have access to the firm's books.

(3) Premium Bonus Schemes. There are several systems of premium bonus schemes. In the building industry they are probably the most commonly used to arrive at incentive payments.

They are based on a different concept from the other incentive schemes in that the employee is paid at ordinary time work rates for the hours worked plus a bonus based on the number of hours saved. The employee may increase his wages but he cannot lose money because a bonus scheme is in operation. It is, therefore, a combination of time work rates and piece work rates.

The amount of time saved that is paid to the employee as bonus varies with the scheme used. This decision is an important one as it affects the setting of targets, the bonus calculation and the method of control.

The more common methods of distributing the time saved are:

(a) 100 per cent scheme. All the time saved is paid to the workmen as bonus.
(b) 50 per cent scheme, or Halsey scheme. The workmen are paid a fixed percentage of the time saved, ie 50 per cent of time saved is paid as bonus.
(c) Rowan scheme. The bonus hours are calculated using the formula:

$$\text{Bonus hours} = \frac{\text{time taken} \times \text{time saved}}{\text{time allowed}}$$

There are other schemes using different percentages paid to the workmen and also curved geared methods.

Geared incentive schemes were introduced into the building industry to allow production payments to be made to workmen for lower performances than with the 100 per cent scheme, particularly where quality is important.

In the 100 per cent scheme there is the psychological advantage that the operatives feel that they are being fairly treated in that they are paid the whole of the saving. Employees working at standard performance will earn the same on this scheme as they would on other schemes, and workmen exceeding standard performance earn higher bonus than on other schemes. It has the disadvantage that, for the slow worker, bonus starts at a higher level of performance than on other schemes (no bonus earned at 75 rating or under) and that management do not get any benefit from the fast worker to help finance the running of the scheme or for work done at under 75 performance rating. The greater amount of work done by the men in the same time, however, increases turnover and reduces the effect of the fixed overheads.

The 50 per cent scheme and the Rowan scheme have the advantage that the employees start earning bonus at a lower performance level and that the employers get the advantage of savings with performances of workmen above standard performance level. At standard performance the workmen will earn the same amount on this scheme as they would on other schemes. There is, therefore, an incentive to both workmen and management to see the scheme functioning properly. It has the disadvantage, however, that the men feel that they are being paid less than they are entitled and that cost control may not be as straightforward as in the 100 per cent scheme.

In order to calculate the bonus hours allowed under the various schemes the time required to do the job at standard performance is determined. To this is added a percentage which will give the workmen a bonus of $33\frac{1}{3}$ per cent of the basic rate.

In the 100 per cent scheme add $33\frac{1}{3}$ per cent.

eg 6 h (standard performance) + $33\frac{1}{3}\%$ = 8 h (target)

In the 50 per cent scheme add $66\frac{2}{3}$ per cent

eg 6 h (standard performance) + $66\frac{2}{3}\%$ = 10 h (target)

The effect of incentive schemes on estimating

The incentive scheme should be based on the outputs allowed for in the estimate. If this is not done then the contractor is in a dangerous position, not knowing whether he is paying his men too high a proportion of money for which he has contracted to do the work. The amount of bonus to be paid to the men must be planned. The increased payments he is making to the men must be allowed for in the estimate. If this simply means an addition on the rates then the principles of a good incentive scheme are not being met and the contractor will be less competitive. Bonus payments should be coupled with greater productivity, and this should be reflected in the labour outputs used to prepare the estimate.

To calculate the cost of labour working at standard performance (ie 100 rating):

	Joiner £	Joiner £
Wage for week of 40 h	60.00*	67.00†
Planned bonus —30%	18.00	
	78.00	67.00
	÷ 40	÷ 40
Hourly rates —with bonus	£1.95 without bonus	£1.68

At standard performance a joiner will lay 10 square metres of 25 mm T & G softwood board flooring in five hours.

Working at a normal performance of 75 rating the time taken for a joiner to lay 10 square metres of flooring would be:

$$\frac{\text{standard time} \times \text{standard rating}}{\text{observed time}} = \frac{5.0 \times 100}{75} = 6.7 \text{ h.}$$

Cost of labour at 100 performance earning bonus
  joiner 5 h at £1.95 = £9.75
Cost of labour at 75 performance not earning bonus
  joiner 6.7 h at £1.68 = £11.26

---

* Standard basic grade A rate of wages plus Joint Board Supplement (£51.60 + £8.40 = £60.00).
† Guaranteed minimum weekly earnings (see page 75)

COMPARING THE 100 PER CENT AND 50 PER CENT INCENTIVE SCHEMES

(a) 100 per cent scheme

Standard hours	Allowed hours (SH + 33⅓%)	Time taken hours	Time saved hours	Bonus hours	Total hours	Rate per hour	Total wages	Effective hourly rate
21	28	17(125R)	11	11	28	1.36	38.08	2.24
21	28	21(100R)	7	7	28	1.36	38.08	1.81
21	28	28(75R)	—	—	28	1.36	38.08	1.36
21	28	35(60R)	—	—	35	1.36	47.60	1.36

(b) 50 per cent scheme

Standard hours	Allowed hours (SH + 66⅔%)	Time taken hours	Time saved hours	Bonus hours	Total hours	Rate per hour	Total wages	Effective hourly rate
21	35	17(125R)	18	9	26	1.36	35.36	2.08
21	35	21(100R)	14	7	28	1.36	38.08	1.81
21	35	28(75R)	7	3.5	31.5	1.36	42.84	1.53
21	35	35(60R)	—	—	35	1.36	47.60	1.36

At 100 per cent performance the workmen earn the same irrespective of which scheme is being operated. The fast worker in the 100 per cent scheme earns more per hour than the fast worker in the 50 per cent scheme, but the slow worker in the 100 per cent scheme earns less than the slow worker in the 50 per cent scheme. In the 100 per cent scheme, provided that 75 performance is achieved, then the contractor is reimbursed the costs allowed in his estimate, but under 75 performance he makes a loss. In the 50 per cent scheme the workman still earns a bonus even if working at under 100 performance, but the cost to the contractor may exceed the amount of money allowed for that operation in the estimate. This is compensated, however, by the fact that workmen above standard performance are not paid as much as they would have been paid if working in the 100 per cent scheme.

Bonus earning distribution

Bonus payments are either calculated and paid out as an average to all men working bonus on the site or calculated for and paid to individual squads of men. Bonus payments are normally allocated as follows:

Journeymen—four shares; labourer—three shares; apprentices (depending on year)—from half share to three shares.

A portion of bonus earnings may also be paid to non-productive workmen, who themselves cannot earn bonus but due to their efforts make it possible for the workmen to earn bonus.

The targets agreed may represent the labour outputs allowed by the estimator when pricing the bill of quantities, but in order to prevent this information being passed on to competitors, the actual outputs are not normally used.

General
Incentive schemes, as mentioned previously, have three main aims:

(a) by increasing efficiency to reduce the cost of building;
(b) to increase individual and collective production;
(c) to provide opportunity for increased earnings.

A good incentive scheme should be easy to understand, it should state clearly the method of payment which should be straightforward to calculate, and it must be seen to be fair by the operatives. All jobs should be accurately defined and the targets stated. The targets should be issued to the operatives prior to the work commencing and once the scheme is in operation then they should not be altered unless by mutual agreement between the workmen and the management. The work should be continuous and measurable and allowances should be made for any delays that are outside the control of the operatives. Bonus payments should be related to individual or group performances and there should be no restrictions on the amount of bonus that can be earned. Bonus earnings are generally calculated weekly. Overtime working should be paid at basic rates and not at time and a half.

The targets in an incentive scheme should be based on standard times. The amount of work that can be reasonably expected from an average operative may be determined by the use of work study techniques. The operation may be method studied at first to ensure that the best method is being used and then time studied to arrive at a standard time. The men should be informed of the outcome of the method study and shown how to improve their output on this basis. The work targets and the amount of bonus payments should be agreed between the contractor and the union for each site prior to the commencement of the work.

The scheme should be drawn up so as to make it possible for the workmen to earn from 20 to 30 per cent above their normal hourly rates.

In the following examples of a geared bonus scheme the amount of saving in money due to the increase in output by the workmen is divided proportionately between the contractor and the workmen. The workmen will receive a bonus payment of two-thirds of the money saved and the remainder will go to the contractor in order that he may be reimbursed for the cost of operating the scheme and also give him some additional profit.

EXAMPLES OF GEARED SCHEME
(1) Squad of two plasterers and one labourer engaged on bonus on a large building. Targets agreed between contractor and union.

Operation	Basic output per man/h
Fixing plasterboard on walls and ceilings	$3\frac{1}{2} m^2$
Fixing metal corner beads	$7\frac{1}{3} m$
2 coats plaster on lath	$1\frac{7}{8} m^2$
2 coats plaster on brick	$1\frac{1}{3} m^2$
3 coats plaster on concrete	$1\frac{1}{4} m^2$

One week's production:

Operation	Quantity	Target	Hours
Plasterboard	$175 m^2$	$\div 3\frac{1}{2} m^2$	50
2 coats plaster on ditto	$60 m^2$	$\div 1\frac{7}{8} m^2$	32
Metal corner beads	$33 m$	$\div 7\frac{1}{3} m$	$4\frac{1}{2}$
2 coats plaster on brick	$60 m^2$	$\div 1\frac{1}{3} m^2$	45
3 coats plaster on concrete	$40 m^2$	$\div 1\frac{1}{4} m^2$	32
			$163\frac{1}{2}$
		Total h worked	120
		Hours saved	$43\frac{1}{2}$

Bonus payment: $43\frac{1}{2}$ h at £1.07* = £46.56

* Two-thirds of basic hourly rate average of two plasterers (or slaters) and one labourer.

72

Allocation of bonus between workmen:

	Hours		Share	Total	Bonus £
Plasterer	40	×	4	160	16.93
Plasterer	40	×	4	160	16.93
Labourer	40	×	3	120	12.70
	120			440	£46.56

(2) Squad of two slaters and one labourer engaged on bonus on housing scheme. Targets agreed between contractor and union.

Operation	Basic output per man/h
Laying 300 × 150 mm slates, including felt	$1\frac{7}{8}\ m^2$
Raking cutting	$5\frac{1}{2}\ m$
Laying and bedding ridging	$4\frac{1}{2}\ m$

One week's production:

Operation	Quantity	Target	Hours
Laying slates	$261\ m^2$	$\div\ 1\frac{7}{8}\ m^2$	139
Raking cutting	$44\ m$	$\div\ 5\frac{1}{2}\ m$	8
Ridging	$54\ m$	$\div\ 4\frac{1}{2}\ m$	12
			159
		Total h worked	120
		Hours saved	39

Bonus payment: 39 h at £1.07* = £41.73

Allocation of bonus between workmen:

	Hours		Share	Total	Bonus £
Slater	40	×	4	160	15.17
Slater	40	×	4	160	15.17
Labourer	40	×	3	120	11.39
	120			440	£41.73

* Two-thirds of basic hourly rate average of two plasterers (or slaters) and one labourer.

CHAPTER VII

LABOUR COSTS

INTRODUCTION
In most unit rates the cost of labour may be less than the cost of materials, yet it is the determination of the labour cost that calls for the most skill on the part of the estimator.

The labour element influences the unit cost in two ways:

(1) The various factors that the estimator adds to the basic wage rate to give the 'all-in' labour rate.

(2) The amount of labour the estimator considers necessary to complete a set task.

The conditions under which labour is employed in the building industry (ie the Working Rules) and the rates of wages, are laid down by different national bodies:

(a) The National Joint Council for the Building Industry.

(b) The Civil Engineering Construction Conciliation Board for Great Britain.

(c) The Signatory Unions to the Mechanical Construction Engineering Agreement and the Engineering Employers' Federation.

(d) The Joint Conciliation Committee of the Heating and Ventilation and Domestic Engineering Industry.

(e) The Joint Industry Board for Plumbing Mechanical Engineering Services in England and Wales.

(f) The National Joint Council for the Laying Side of the Mastic Asphalt Industry.

(g) The Scottish Regional Committee and by the National Joint Council for the Building Industry.

(h) The Scottish Joint Industry Board for the Electrical Contracting Industry.

(i) The Scottish and Northern Ireland Joint Industry Board for the Plumbing Industry.

The following text is based on the 'National Working Rules for the Building Industry' as laid down by the National Joint Council for the Building Industry:

74

The 'all-in' labour rate
This is the basic wage rate plus some or all of the following:
  (i)   Joint board supplement.
  (ii)   Guaranteed minimum bonus payment.
  (iii)   Travelling time and fares.
  (iv)   Lodging allowance.
  (v)   Overtime.
  (vi)   Supervision.
  (vii)   National Insurance.
  (viii) Holidays with pay.
  (ix)   Redundancy and sundry costs.
  (x)   Construction Industry Training Board Levy.
  (xi)   Extra payments under National Working Rule 3.
  (xxi)   Employers' liability and third party insurance.
  (xiii) Guaranteed time due to inclement weather.

General
If a contract is firm in respect of labour the contractor may require to add an allowance to cover any increase in the labour rate, depending on the size and expected duration of the contract.

The wage rate is reviewed annually by the National Joint Council for the Building Industry. The guaranteed minimum weekly earnings as at June 1979 are as follows:

Craft operatives (Grade A)	£67.00
Labourers (Grade A)	£57.20

These guaranteed minimum weekly earnings are made up as follows:

	Craft operatives	Labourers
	£	£
Standard Basic Grade A		
Rates of wages	51.60	44.00
Joint board supplement	8.40	7.20
Guaranteed minimum bonus payment	7.00	6.00
	£67.00	£57.20

The actual rates of wages do not affect the method used to build up the unit rates. For this reason the rates of wages used in this chapter are:

75

Craft operatives   £70.00 per week
Labourers   £60.00 per week

The normal working week has been taken as 40 hours, which are worked in 5 days, 8 hours per day, Monday to Friday.

The National Joint Council for the Building Industry lay down the working rules for the industry. There are also statutory requirements for National Insurance and the like, which must be taken account of when determining the total labour costs. These rules and allowances vary from time to time, but this will not invalidate the method used for calculating their effect on labour costs.

Before proceeding to worked out labour rate examples incorporating the various points that have been discussed, these rules and allowances will be considered in greater detail.

(i) Joint board supplement
All workmen are entitled to receive a joint board supplement and at present time this amounts to £8.40 per week for tradesmen and £7.20 per week for labourers.

(ii) Guaranteed minimum bonus payment
All workmen are entitled to receive a minimum bonus payment and at the present time this amounts to £7.00 per week for tradesmen and £6.00 per week for labourers.

If an incentive scheme or productivity agreement is in operation in which the bonus earnings fall below the level of the guaranteed minimum for reason beyond the control of the workmen, then the guaranteed minimum bonus must be paid.

Where no incentive scheme or productivity agreement is in operation a sum equal to the guaranteed minimum bonus must be paid to the workmen.

(iii) Travelling time and fares
Where a workman is required to travel daily the distance shall be measured on an Ordance Survey Map 1:50000 Series in a straight line from the centre of the kilometre square in which his home is to the centre of the kilometre square in which lies the place where his employer requires him to work.

Where the distance between the two centres does not exceed six kilometres (3.73 miles) then no travelling or fare allowance shall be paid.

Where the distance between the two centres exceeds six kilometres the workman shall be paid for that distance (one way) a daily travelling allowance of 6p for each kilometre (0.62 mile) or part thereof and a daily fare allowance per kilometre or part thereof in accordance with the following scale:

Distance (km)	Allowance
1–6	Nil
7	9p
8	17p
9	25p
10	32p
11	38p
12	44p
13	50p
14	55p
15*	59p
20	84p
30	£1.34
40	£1.84
50	£2.34

This rule does not apply for distances exceeding 50 kilometres (31.05 miles). Time spent on daily travelling is deemed to be outwith the working day.

(iv) Lodging allowance

Calculated on the basis of the working rule agreement. The allowance at present is £3.75 per night on which lodgings are required.

An operative sent to a distant job to which he does not travel daily shall be entitled to the payment of fares or conveyance in his employer's transport as follows:

(i) From his home to the job at commencement and from the job to his home on completion, and

(ii) From the job to his home and back at periodic leave intervals related to the straight-line distance between the job and his home on the following basis:

* Rates are also given in the National Working rates for each 1 kilometre distance between 6 and 50 kilometres.

50 km and up to and including 80 km—two weeks.
Over 80 km and up to and including 120 km—three weeks.
Over 120 km—at intervals fixed by mutual agreement between the operative and his employer before the operative goes to the job.
Travelling time allowances are set out in the working rule agreement.

(v) Overtime
Calculated on the basis of the working rule agreement. Overtime is paid at the following rates for a 5-day week.
Mondays to Fridays:
First three hours, time and a half; afterwards, until starting time next morning, double time.
Saturdays and Sundays:
Time worked from starting time on Saturday morning until 4 pm, time and a half; from 4 pm on Saturday until normal starting time on Monday morning, double time.
Holidays:
For recognised public holidays, double time.
For local holidays, time and a half for the period of the normal working day and double time thereafter.

(vi) Supervision
Generally all supervision which is permanently on the job is charged against the job. This may range from an agent with general foreman, time-keepers, etc, to a working foreman on the job. Supervision by a travelling foreman who is responsible for several jobs should be covered by the general overheads. The total cost of supervision per week should be divided by the average weekly man-hours and included in the labour cost.

(vii) National Insurance
National Insurance contributions are related to earnings and are calculated on gross pay. There are two schemes (1) not-contracted out, and (2) contracted

out. The former scheme makes provision for a state pension while in the latter the employer must operate a pension scheme which meets the requirements laid down in the Social Security Pensions Act 1975.

Contributions are calculated either by (a) applying the prescribed percentages to gross pay, or (b) by referring to National Insurance contribution tables.

In the following examples the not-contracted out scheme has been used and 13.5 per cent applied to gross wages in order to arrive at the employer's contribution. The maximum contribution per week is £16.20.

(viii) Holidays with pay

Annual
The value of holiday credit stamps as at August 1979 is as follows:

Adults	£7.70*
Under 18	£5.70

For the purpose of the 'all-in' rate the employer's contribution for annual holiday stamps has been taken as £8.00 per week or 49 weeks at £8.00 = £392 per annum.

Public
Local and public holidays have been taken as amounting to the equivalent of $1\frac{2}{5}$ weeks per annum. With a wage-rate of £67.00 per week then the equivalent cost would be:

£93.80 per annum or
£1.91 per week.

For the purposes of the 'all-in' rate the allowance for public holidays has been taken as £2.00 per week or £100.00 per annum.

(ix) Redundancy and sundry costs
An allowance must be made to cover sickness benefit, severance pay, loss of production during notice, absenteeism, and the like. The allowance is normally

* This includes a contribution of 10p towards the death benefit scheme.

made as a percentage of the wages paid and will vary according to the type and experience of the firm. Generally it is in the region of one per cent.

(x) Construction Industry Training Board Levy
The employer must pay an annual levy to the Industrial Training Board to cover the cost of the apprenticeship training scheme. Money is refunded to the firm if their apprentices attend approved educational establishments.
The cost to the firm will vary, but will be in the region of £35 per annum for tradesmen and £6 per annum for labourers (ie £0.88/h and £0.15/h respectively). Alternatively, this may be considered as an overhead and charged for as explained in Chapter IV.

(xi) Extra payments under National Working Rule 3
Allowances for tool money, dirty money, and the like, are laid down in the working rules agreement, eg carpenters and joiners providing and maintaining own tools receive £0.50 per week tool allowance, and extra payment for work in water or close contact with dirt or filth, £0.04 per hour.

(xii) Employer's liability and third party insurance
Premium is generally calculated as a percentage of the wage bill. The amount would vary with the trade and type of work carried out by the firm.
Allow a premium of £1.75 or £2.25 per £100.00 of the total wage bill. In this calculation the amount paid as wages includes overtime and travelling time but does not include allowances for fares, etc, paid with wages.

(xiii) Guaranteed time due to inclement weather
The employee is paid forty times the hourly wage rate applicable to him, ie the normal 40 hour week is fully guaranteed. The contractor, therefore, must allow for excessive stoppages due to inclement weather. The percentage allowed will depend on the type of work to be undertaken and the time of year. The percentage normally varies from nil to 10 per cent.

LABOUR RATE EXAMPLES

Examples 1 to 5 are computed on an hourly basis, while Examples 6 to 8 are on an annual basis.

(1) To calculate the 'all-in' hourly labour rate for tradesmen and labourers working under normal conditions.

	Tradesman £	Labourer £
Wage at standard basic rate, joint board supplement, guaranteed minimum bonus: 40 h	*70.00*	*60.00*
*Supervision: 40 h at £0.15 per h for working foreman: £6.00 divided by 11 men	*0.55*	*0.55*
	70.55	*60.55*
National Insurance 13.5%	*9.52*	*8.17*
Annual and public holidays	*10.00*	*10.00*
Redundancy and sundry costs 1% on £70.55 and £60.55	*0.71*	*0.61*
CITB levy	*0.88*	*0.15*
†Tool allowance (joiner)	*0.50*	*—*
	92.16	*79.48*
Employer's liability and third party insurance 2%	*1.84*	*1.59*
	94.00	*81.07*
Guaranteed time 3%	*2.82*	*2.43*
	40)96.82	*83.50*
	£2.42	*£2.09*

(2) To calculate the 'all-in' hourly labour rates for tradesmen and labourers working under normal conditions, but on a site 30 km distant (measured in accordance with the working rule agreement), and travelling by public transport.

* Supervision: An average gang size has been assumed which comprises a working foreman, six tradesmen and four labourers.

† The tool allowance varies with different trades. Amounts are laid down in the working rule agreement.

		Tradesman £	Labourer £
Wage as standard basic rate, joint board supplement, guaranteed minimum bonus: 40 h		70.00	60.00
Travelling allowance 5 days × 30 km × £0.06		9.00	9.00
Supervision:			
Extra payment 40 h at £0.15 per h	= £6.00		
Travelling allowance 5 days × 30 km × 6p	= £9.00		
	£15.00		
Divide £15.00 by 11 men gang		1.36	1.36
		80.36	70.36
National Insurance 13.5%		10.85	9.50
Annual and public holidays		10.00	10.00
Redundancy and sundry costs 1% on £80.36 and £70.36		0.80	0.70
CITB levy		0.88	0.15
Tool allowance		0.50	—
Fare allowance 5 days × £1.34		6.70	6.70
		110.09	97.41
Employer's liability and third party insurance 2%		2.20	1.95
		112.29	99.36
Guaranteed time 5%		5.61	4.97
		40)117.90	104.33
		£2.95	£2.61

(3) To calculate the 'all-in' hourly labour rates for tradesmen and labourers working under normal conditions but on a site 10 km distant (measured in accordance with the working rule agreement), and transport is supplied by contractor.

Contractor to supply lorry to transport the men to the site, return to the yard for general use and return for men at night.

Morning run: 12 km out and 12 km back (actual distance and not straight line measurement). 24 km in all and allowing for delay on site and for incidental delays, say 1 h.

Every run takes 1 h
Lorry time 2 h per day
5 days × 2 h = 10 h per week
Cost of lorry (hire rate)   *£2.00 per h*
Driver (labourer rate* + £0.12) £2.82 per h

       £4.82 per h or £48.20 per
           week

With a gang of 11 men the cost would be £4.38 per man per week.

	Tradesman £	Labourer £
Wage at standard basic rate, joint board supplement, guaranteed minimum bonus: 40 h	70.00	60.00
Travelling allowance 5 days ×10 km ×£0.06	3.00	3.00
Supervision:		
Extra payment		
40 h at £0.15 per h = £6.00		
Travelling allowance = £3.00		
£9.00		
Divide £9.00 by 11 men	0.82	0.82
	73.82	63.82
National Insurance 13.5%	9.97	8.62
Annual and public holidays	10.00	10.00
Redundancy and sundry costs 1% on £73.82 and £63.82	0.74	0.64
CITB levy	0.88	0.15
Tool allowance	0.50	—
Cost of transport	4.38	4.38
Carried forward	100.29	87.61

* See note on page 91.

		Brought forward	100.29	87.61
Employer's liability and third party				
	insurance 2%		2.01	1.75
			102.30	89.36
Guaranteed time 5%			5.12	4.47
			40)107.42	93.83
			£2.69	£2.35

(4) To calculate the 'all-in' hourly rates for tradesmen and labourers working under normal conditions. The men to work overtime 3 h per day from Monday to Thursday.

		Tradesman £	Labourer £
40	*Wage at standard basic rate, joint board supplement, guaranteed minimum bonus:*		
	40 h	70.00	60.00
	Overtime 3 h × 1 ½ = 4½h		
18	*4½h × 4 days = 18 h*		
—	*at £1.75 and £1.50*	31.50	27.00
58	*Supervision: 58 h at £0.15*		
	per h £8.70 ÷ 11 men	0.79	0.79
		102.29	87.79
National Insurance 13.5%		13.81	11.85
Annual and public holidays		10.00	10.00
Redundancy and sundry costs			
1% on £102.29 and £87.79		1.02	0.88
CITB levy		0.88	0.15
Tool allowance (bricklayer)		0.25	—
		128.25	110.67
Employer's liability and third party			
insurance 2%		2.57	2.21
		130.82	112.88
Guaranteed time 5%		6.54	5.64
	Carried forward	52)137.36	118.52

	Brought forward	*52)137.36*	*118.52*
Net number of hours worked:			
40 + (3 h × 4 days)			
40 +12 = 52 h		*£2.64*	*£2.28*

(5) To calculate the 'all-in' hourly labour rates for tradesmen and labourers for the following conditions. Site is 60 km distant (measured in accordance with the working rule agreement), and the men are paid lodging allowance. The majority of the work will be carried out in conditions unaffected by the weather.

Cost of return fare is £5.00

Hours worked: 8 am to 8 pm weekdays; 8 am to 12 noon Saturdays

Gang consists of 1 working foreman, 10 tradesmen and 10 labourers

			Tradesman £	Labourer £
40	*Wage at standard basic rate, joint board supplement, guaranteed minimum bonus:*			
	40 h		*70.00*	*60.00*
	Overtime			
	3 h × 1½ = 4½ h × 5 days =	*22½*		
	4 h × 1½ = 6 h × 1 day =	*6*		
		28½		
28½	*28½ h at £1.75 and £1.50*		*49.88*	*42.75*
1	*Travelling time 2 h per fortnight*			
	1 h/week at £1.75 and £1.50		*1.75*	*1.50*
69½	*Supervision: 69½ h at £0.15 per h*			
	£10.43 ÷ 11 men		*0.95*	*0.95*
			122.58	*105.20*
	National Insurance 13.5%		*16.55*	*14.20*
	Annual and public holidays		*10.00*	*10.00*
	Carried forward		*149.13*	*129.40*

	£	£
Brought forward	149.13	129.40
Redundancy and sundry costs		
1% on £122.58 and £105.20	1.23	1.05
CITB levy	0.88	0.15
Tool allowance (joiner)	0.50	—
Lodging allowance: 7 nights at		
£3.75	26.25	26.25
Fares: £5.00 per fortnight (per week)	2.50	2.50
	179.49	159.35
Employer's liability and third party		
insurance 2%	3.59	3.19
	59)183.08	162.54
Net number of hours worked:		
40 + (3 × 5) + (4 × 1)		
40 + 15 + 4 = 59 h	£3.10	£2.75

(6) To calculate the 'all-in' labour rates for apprentices.

A second year apprentices' rate has been taken as an average rate. The rate has been taken, according to the working rule agreement, as 80 per cent of the tradesman's basic rate plus joint board supplement, plus guaranteed minimum bonus.

	£
Wage per year = 52 weeks at £56.00	2 912.00
National Insurance	394.34
	3 306.34
Employer's liability and third party	
insurance 1.50%	49.60
Cost of second year apprentice per year	£3 355.94
Total number of working days per year	
= 52 weeks × 40 h	= 2 080 h
Carried forward	2 080

		£
	Brought forward	2 080

Deduct
Annual holiday: 3 weeks × 40 h	= 120	
Public holidays: 7 days × 8 h	= 56	
Day release classes: 44 days × 8 h	= 352	
Sickness, say average 4 days × 8 h	= 32	560

1 520 h

Actual number of hours worked = 1 520

$$Cost\ per\ hour = \frac{£3\ 355.94}{1520} = £2.21$$

(7) Computation of labour rates on an annual basis.

Total number of hours worked per year
Standard time 52 weeks × 40 h		= 2 080
Overtime (average)		200

2 280

Less
Annual holidays: 3 weeks × 40 h	= 120	
Public holidays: 7 days × 8 h	= 56	
Sickness (average): 4 days × 8 h	= 32	

208

Actual number of hours worked   2 072

	Tradesman £	Labourer £
Wage at standard basic rate		
2072 h at £1.29 and £1.10	2 672.88	2 279.20
Joint board supplement		
48 w at £8.40 and £7.20	403.20	345.60
Guaranteed minimum bonus		
48 w at £7.00 and £6.00	336.00	288.00
Non-production overtime		
100 h at £1.29 and £1.10	129.00	110.00
Supervision (2072 + 100) =		
2172 h at £0.15 divided by		
11 men	29.62	29.62
Carried forward	3 570.70	3 052.42

Brought forward	*3 570.70*	*3 052.42*
Annual and public holidays	*492.00*	*492.00*
	4 062.70	*3 544.47*
National Insurance 13.5%	*548.46*	*478.50*
Redundancy and sundry costs		
1% on £4 062.70 and £3 544.42	*40.63*	*35.44*
CITB levy 1 year	*35.00*	*6.00*
Tool allowance (joiner)		
48 w at £0.50	*24.00*	*—*
	4 710.79	*4 064.36*
Employer's liability and third		
party insurance 2%	*94.22*	*81.29*
	4 805.01	*4 145.65*
Guaranteed time 3%	*144.15*	*124.37*
	2 072)4 949.16	*4 270.02*
'All-in' hourly labour rates	*£2.39*	*£2.06*

(8) Computation of labour rate on an annual basis for an advanced plumber

Total number of hours worked per year		
Standard time 52 weeks × 38 hours[1]		*1 976*
Overtime (average)		*300*
		2 276
Less		
Annual holidays: 3 weeks × 38 hours	*114*	
Public holidays: 7 days × 7.6 hours	*54*	
Sickness (average): 5 days × 7.6 hours	*38*	
		206
Actual number of hours worked		*2 070*

	Advanced Plumber
	£
Wage as standard basic rate[2]	
2070 h at £2.24	*4 636.80*
Non-productive overtime[3]	
80 h at £2.24	*179.20*
Public holidays with pay	
7 days × 7.6 h at £2.24	*119.17*
Carried forward	*4 935.17*

88

Brought forward	*4 935.17*
Annual holidays with pay and	
sickness benefit[4]	
49 w at £8.11	*397.39*
	5 332.56
National Insurance[5]	*522.93*
Industrial pension scheme[5]	*359.95*
Redundancy and sundry costs 1%	*53.33*
CITB levy	*60.00*
Tool money 48 w at 50p	*24.00*
	6 352.77
Employer's liability and third party	
insurance 2%	*127.56*
	6 479.83
Guaranteed time 2%	*129.60*
	2072) £6 609.43

'All-in' hourly rate for an advanced plumber £3.19

Note:

[1] *Plumbers' working week reduced to 38 hours in February 1980.*

[2] *Wage rate as at February 1980.*

[3] *Overtime premium rates are only paid after 42 hours are worked per week.*

[4] *Stamp values at October 1979.*

[5] *National Insurance*

Plumbers are contracted out of the state pension scheme and their National Insurance contributions are therefore calculated as follows: 13.5 per cent on the first £19.50 and 9 per cent on the remainder between £19.50 and £135.00 per week, based on 49 weeks.

In order to cover the cost of their industry's pension scheme a charge equivalent to 6.75 per cent of gross wages must be made.

Alternative method

It has been shown in the previous examples how to build up an 'all-in' labour rate for specific jobs and how to make allowances for any requirements that are considered necessary. Most contractors, however, carry out contracts within certain limited areas and in roughly similar conditions. These contractors might find it more convenient to calculate a basic 'all-in' labour rate which only includes for the basic wage, National Insurance contributions and holidays with pay. Additional indirect charges on wages, such as guaranteed time, non-productive overtime, sickness, travelling time and expenses, CITB levy and insurances, can each be determined from past records and charged for as a percentage of the labour costs. Since these indirects are based on previous records, and most contracts undertaken are similar in nature, then the estimator has a greater control over the recovery of these monies. When pricing a contract which is different from the normal pattern of contracts undertaken, then any peculiarities, such as overtime or travelling time beyond the normal allowances, must be ascertained. The anticipated additional expenditure for these items should be calculated and added as a percentage or as a lump sum to the contract. The following is an example of this method:

	Tradesman £	Labourer £
Wage at standard basic rate, joint board supplement, guaranteed minimum bonus: 46 h	*80.50*	*69.00*
Supervision: 46 h at £0.15 per h		
£6.90 ÷ by 11 men	*0.63*	*0.63*
	81.13	*69.63*
National Insurance	*10.95*	*9.40*
Annual and public holidays	*10.00*	*10.00*
Tool allowance	*0.25*	*—*
	46)102.33	*89.03*
	£2.22	*£1.94*

The men are paid for a 40 h week but in addition they work 6 h overtime per week. This makes the number of hours worked per week 46 and the number of hours paid for 49. The non-productive overtime is allowed for as a percentage addition on labour in the additional indirects.

The cost of additional indirects on labour assessed from previous records:

	Percentage of £1.00 of labour costs
Guaranteed time	3.5
Non-productive overtime	5.2
Sickness	0.5
Travelling time and expenses	3.0
CITB levy	0.8
Insurances	1.5
	14.5%

General

As previously mentioned, unit rate build-ups have been based on the labour rate (Example 1) for tradesmen and labourers (irrespective of the type of tradesmen). The 'all-in' rate for apprentices has been taken from Example 6. An example using an alternative method of calculating labour rates is given in Chapter XIII (Brickwork and Blockwork) page 154.

Note: For the purpose of this book the following 'all-in' labour rates have been adopted for the building up of unit rates.

Tradesmen	= £3.00/h
Labourers	= £2.70/h
Apprentices	= £2.50/h

CHAPTER VIII

MOTOR TRANSPORT

Labour and materials moved by the contractor are usually moved by private road haulier or by his own lorries.

If motor transport is only required for short, intermittent periods it is usually more economical for a contractor to use hired transport than to keep his own lorries. It is usually found convenient to own and use one lorry or a small fleet of lorries, depending on size and type of work undertaken. But for any additional demands for such service the work would be carried out by hiring.

Cost of operating builder's own lorry
Cost per hour, excluding normal business overheads and garaging, based on the following:

  800 km per week.
  Capital cost of £7 000.00 with residue value of £800.00 after 5 yr.
  5 km per litre of diesel.
  New tyres every 40 000 km. Tyres cost £500.00 per set.
  Major overhaul every 80 000 km. Overhaul costs approx. £600.00.
  General maintenance, oil and renewals approx. £700.00 per annum.

		£
Standing charge per week		
Driver's wages, 40 h at £2.82*		*112.80*

	£
Capital expenditure	7 000.00
Less residue value†	800.00
	£6 200.00

Carried forward *112.80*

* A driver is paid £0.12 more than a labourer. If lorry driver is also a labourer then he would not be included in the standing charges if the lorry was not in use.

† The residue value need not be deducted. See note on replacement of mechanical equipment in next chapter.

	Brought forward	*112.80*
Annual sinking fund —20% of £6 200.00		
÷ 49 weeks		25.31
Interest on capital —8% of £7 000.00		
÷ 49 weeks		11.43

	£	
Annual road tax	150.00	
Annual insurance	250.00	

£400.00 ÷ 49 weeks   8.16

Running costs per week

Diesel, 800 km ÷ 5 km at £0.20 per litre	32.00
Tyres, £500.00 every 40 000 km, ie £500 per yr ÷ 49 weeks	10.20
Overhaul, £600.00 every 80 000 km, ie £300 per yr ÷ 49 weeks	6.12
General maintenance, £700.00 per yr ÷ 49 weeks	14.29

Standing charges and running costs per week
based on 800 km   £220.31

$$\text{Cost per h} = \frac{£220.31}{40} = £5.51$$

Capacity of motor lorries

Load capacities are somewhat nominal and vary considerably, depending on the manufacture of the lorry.

The load to be carried may be restricted by the weight capacity or by the size of the box. If the capacity is in doubt it is best to calculate the box capacity and convert the cubic capacity into kilograms, bricks, etc.

The box sizes of typical lorries are approximately:

2 740 × 1 830 × 450 mm
2 450 × 2 130 × 450 mm
3 350 × 2 130 × 600 mm

EXAMPLES
(1) Calculate the cost of removing excavations to a tip 1 km distant, using contractor's own lorry.

93

Lorry operating cost = £5.42 per h (from previous example).

	m^3
Box size 3.35 × 2.13 × 0.60	*4.28*
Add for load being above lorry sides (say) 20%	*0.84*
	5.12

A 5.12 m³ capacity of loose soil would correspond to 5.12 m³ less 25% for soil before excavating. 5.12 m³—25% = 5.12 — 1.28 m³. Allow 3.75 m³.

Lorry time to load	h
Three labourers loading*	*0.75*
Travelling and tipping	*0.50*
	1.25

	£
Lorry costs, 1.25 h at £5.51	*6.89*
Labour costs, 2.25 h at £2.70	*6.08*
	£12.97

Cost of removing excavations

$$per\ m^3 = \frac{£12.97}{3.75\ m^3} = £3.46$$

(2) Calculate the cost of transporting bricks per 1000 from the contractor's yard to site 25 km distant. Hire charge of lorry, including driver, is £6.00 per h.

Assume lorry will carry bricks to site and return empty.

	m^3
Capacity of lorry, 2.45 × 2.13 × 0.45	*2.35*
Add for lorry being stacked up higher than sides of box (say) 25%	*0.59*
	2.94

---

* Alternatively, mechanical loading could be used.

*1 m³ contains approx 500 bricks (theoretical maximum)
2.94 m³ contain 2.94 × 500 = 1470 bricks (theoretical
capacity of lorry).*

*The actual capacity would vary according to how the lorry
was loaded, the weight it was capable of pulling, and the type of
roads or site over which it had to travel. Assume the actual
capacity as 1200 bricks.*

Lorry time	*h*
Waiting while bricks are loaded	*0.35*
Travelling to site	*0.85*
Unloading (tipping)	*0.15*
Returning empty from site	*0.75*
	2.10

	£
Cost of lorry transporting, 1200 bricks	
= 2.10 h × £6.00	*12.60*
Labour costs, 4 labourers 0.35 h each	
= 1.40 h × £2.70	*3.78*
Cost of lorry and labour transporting 1200 bricks	*£16.38*

Cost of transporting 1000 bricks = £13.65

*Facing bricks would require labour for unloading at the site
as they should not be tipped.*

*(3) Calculate the cost of transporting slates per 1000 from the
railway station to the site 15 km distant. Hire charge of 6 tonne
lorry, including driver, is £6.00 per h. The slates weigh 150.0
kg per 1000.*
The number of slates per load = 4000.

Lorry time

	h
Travel to station	0.5
Load lorry: 4 labourers 2 h	2.0
Travel to site	0.75
Unload and stack	2.0
	5.25 h

	£
Cost of lorry transporting 4000 slates	
= 5.25 h × £6.00	31.50
Labour costs, 4 labourers 2 h = 8 h × £2.70	21.60
Cost of lorry and labour transporting 4000 slates	£53.10

Cost of transport 1000 slates = £13.28

CHAPTER IX

MECHANICAL EQUIPMENT

INTRODUCTION

There are two main classifications of plant: (a) equipment used on the site and (b) plant used in the workshop. The type of plant dealt with here is the plant used on the site and the general notes that follow are applicable to excavator equipment, concrete mixing plant, and the like.

Plant used in the workshop is static, housed in good conditions and generally used to produce standard or similar items. It is not affected by site and weather conditions. The most usual types are joiners' machinery for making doors, and so on, and equipment for the manufacture of precast concrete units.

Generally plant is employed on a building site in order to save money, labour or time, or a combination of all three. In many circumstances it is cheaper to use machines for certain operations than to use the corresponding amount of labour, for example for large site excavations. With the scarcity of building labour the contractor must become more mechanised in order to cut down on the amount of labour required. In certain circumstances he may require to do this irrespective of cost. In contracts where time is an important factor the contractor may require to use certain plant in order to speed up the work and get it completed within the required time.

For pricing purposes plant used on the site can be subdivided into (a) plant performing specific items of work and (b) plant which performs many different tasks.

COST OF PLANT

The cost of plant can be considered under the main headings of (a) standing charges, (b) running costs and (c) variable costs. These headings can be further detailed as follows:

(1) Capital outlay	
(2) Replacement	Standing charges.
(3) Maintenance and repairs	
(4) Labour operating and attending	Running costs.
(5) Fuel	
(6) Transport	Variable costs.
(7) Temporary site work	

All these items must be considered and broken down into the cost per unit of measurement which is required to be priced or into the cost per week in order to arrive at a total charge for the contract.

STANDING CHARGES

The standing charges are irrespective of the cost of running the machine, and the more the machine is used the less these charges influence the daily or hourly cost of operating the machine. The actual working time per annum is estimated and the charge per hour calculated in order to recover the cost of the standing charges in the rate charged. The points to be considered in each item are as follows:

Capital outlay

The purchase of a machine should be considered as an investment. Interest on the capital outlay would therefore be expected. An interest charge compatible with other similar investments would be in the region of 5 to 10 per cent.

Replacement

The machine requires to earn sufficient money to buy a similar machine at the end of its useful working life. The two main points to be considered are: (a) the useful life of the machine and (b) the expected cost of the replacement machine. The useful life depends on the type of machine and use to which it is put. An excavator would be expected to have a shorter life than a crane. If a machine has a useful life of five years then a 20 per cent return on capital cost would be required. For a life of ten years a 10 per cent return would be required. The expected cost of the replacement machine may be greater than the cost of the machine it is replacing. This is, however, balanced to some extent by the market or scrap value of the machine to be replaced and can generally be discounted.

Maintenance and repairs

This item deals only with the labour and material costs of periodic services and replacement of worn parts. It does not take into account any time lost due to these services or breakdowns. The cost involved depends on the type of equipment and can vary from about 5 per cent to 50 per cent. The actual percentage is calculated from previous records kept by the individual contracting firms.

RUNNING COSTS

Running costs are only incurred when the machine is actually being used. As with the standing charges, the actual work time per annum is estimated and the charge per hour calculated. The points to be considered in each item are as follows:

Labour operating and attending

This item must cover the cost of all labour necessary for operating the machine. In the case of a concrete mixer this would be the operator and the attendant labourers for aggregate and cement; and for excavators and cranes, drivers and banksmen would be required. The labour removing the concrete to the required position would not be included in this item as this is over and above the mixing which the machine is expected to perform.

Fuel

This item covers the cost of all fuel, oil, etc, required for the efficient running of the machine.

VARIABLE COSTS

The total of the standing charges and running costs make up the basic cost of operating the machine. These costs are independent of the site and remain constant. Costs which vary from site to site must be considered separately and are as follows:

Transport

The cost of transporting the machine on and off the site must be considered. Generally the cost of transport to the site is charged against the contract and the cost of transport off the site is charged against the contract at which the machine is required. The cost of transport is usually calculated as a lump sum and charged for under the preliminaries section of the bill of quantities.

Temporary site work

Any additional work which is required due to the site conditions is charged for under this heading. For example, an opening may be required to give access to the site for the machine which would have to be made good at completion, or a hardstanding may be required to be built for the concreting plant which would have to be removed at completion of the contract. This would also be calculated as a lump sum and charged for under the preliminaries section of the bill of quantities.

Calculation of time worked by machine

The machine will not work every day throughout the year and it is therefore necessary to calculate the actual number of possible working days:

		days
Days not worked:	Weekends	104
	Annual holidays	15
	Public holidays (approx.)	7
		126

Time will also be lost in transporting equipment from site to site and in idle time while waiting to be transported. This may account for a further 25 to 35 days of non-productive time.

In addition, time is also lost due to overhauls and repairs. The amount of time wasted is influenced by the type and use of the machine and may account for another 10 to 20 days of non-productive time.

This gives a total of about 160 to 180 non-productive days or 185 to 205 productive days.

It must also be considered that during the normal working day there are delays and stoppages and that it is impracticable to work continuously throughout the whole day.

PLANT HIRING

The types of plant used on a building site will vary from small hand tools, such as electric drills, to pneumatic drills and concrete mixers, to large earth-moving equipment and tower cranes.

As mentioned previously, if a contractor cannot find continuous use for mechanical plant and it is likely to stand idle for long periods, then it will be more economical to hire the plant for the specific job or operation. The high capital cost of many items of plant and specialisation of design leading to particular requirements, has led to an increasing use of plant hire. The plant hire sector of the construction industry is therefore firmly established and has shown steady growth over the last decade.

Large contracting organisations, who own plant, usually have a plant pool and a separate plant department which hires internally to the contracting department or externally to other contractors, if it is not required by its own organisation, at rates similar to those normally charged by plant-hiring firms. By this means they achieve better control over the recovery of the costs of the plant.

The main advantages of hiring are:

(a) Increase in liquidity. Large capital outlays could affect the firm's cash flow and may inhibit other forms of investment within the business. Money spent on plant and machinery cannot be used to expand the business, pay wages to employ more men and increase turnover, or pay invoices more quickly so as to get better discounts.

(b) Greater efficiency. The contractor can hire the most efficient plant for the job, whereas his own plant may be adequate for a whole range of work but not ideally suited for particular needs. Operators from plant-hire firms are trained in the use of the machine and therefore are probably safer and more efficient than operators on machines which are less satisfactory for the particular work.

(c) Obsolescence. Plant is being continually improved with new models being introduced. The contractor could be left with outdated plant which has little value other than for scrap even if it had been well maintained and under-utilised.

(d) Greater control of costs. Hire costs are known at the

101

time of estimating and can be fully covered in the tender figure. Extra costs, over and above the hire charge, such as insurance, are known and the appropriate allowance made. With the contractor's own plant, costs and charges could vary due to unexpected breakdown or higher than normal maintenance costs.

The main disadvantages of hiring are:

(a) Availability. Owned plant is under the control of the contractor and can be made available when required. Suitable plant may be unobtainable from plant-hire firms at the dates which suit a particular contract programme. The removal of this uncertainty may assist in contract planning.

(b) Loss of grants. In certain circumstances a contractor may receive an investment grant to aid in the purchase of plant and machinery. Such grants will not be obtainable if the plant is hired.

Whether to buy or hire plant is an important company decision. It is necessary to predict future workflow in order to assess the level of plant requirement which should be purchased as against hired. This policy decision can be reviewed periodically, but it is possibly unwise for a firm to either own or hire all its plant requirement.

EXAMPLES

(1) Cost of excavations per m³ using D6 tractor and scraper of 4.5 m³ capacity which belongs to the contractor.

This type of plant consists of two units: the tractor and the scraper. It provides a cheap form of excavating if the haul does not exceed 450 m. Beyond this distance the mechanical excavation plant should be used in conjunction with vehicular transport. This type of machine is most suited for large road contracts or site levelling work. The scraper acts as a container in which the excavated material is hauled.

As the scraper is pulled by the tractor it excavates the ground with its cutting edge and the material enters the scraper which, when full, is raised above the ground and hauled to the tip by the tractor for off loading.

As mentioned in the introduction to this chapter, a major

consideration to be determined before purchasing a machine is its capacity. This is not what it can achieve in one hour but what it can achieve in one year, allowing for the weather, repairs, holidays, waiting time between jobs and time lost in transporting from job to job.

Cost of machine

		£
Capital cost		*75 000.00*
Replacement—5 yr	*20%*	
Interest	*10%*	
Repairs	*25%*	
	55%	

∴ Total annual cost = £41 250.00
Use 150 days = £275.00 per day or £34.38 per h.

Cost of using the machine to excavate and remove to spoil heaps not exceeding 100 m.

	£
Cost of tractor and scraper per h	*34.38*

Labour operating and attending

	£	
Driver (1)	*2.87*	
Banksman (1)	*2.70*	
Labourers (2)	*5.40*	
		10.97

Fuel

	£	
Diesel, 20 litre at £0.15	*3.00*	
Oil and grease	*0.60*	
		3.60
		£48.95

Productive rate
4.5 m³ scraper at 100 m haul would give 8 turns per h.
36 m³ excavated and deposited.

$$Cost\ per\ m^3 = \frac{£48.95}{36\ m^3} = \qquad \begin{array}{c}£\\1.36\end{array}$$

Profit and oncost 15%	0.20
	£1.56 m³

Cost per m³—£1.56

(2) In Example 1, if instead of owning the D6 tractor and scraper, the contractor hired one for the excavations, the build-up of rate would have been as follows:
In this case the capital outlay, replacement, and maintenance costs are all covered by the hire charge.

Cost of machine

£

Hire charge per h — 36.00

Labour operating and attending

£

Driver (1)	2.87
Banksman (1)	2.70
Labourers (2)	5.40
	10.97

Fuel

Diesel, 20 litre at £0.15	3.00
Oil and grease	0.60
	3.60
	£50.57

Productive rate
As before 36 m³ excavated and deposited.

$$Cost\ per\ m^3 = \frac{£50.57}{36\ m^3} = £1.40$$

Profit and oncost 15%	0.21
	£1.61 m³

Although in this example it is dearer to hire, this is only true if the equipment owned is used to its full capacity. If the equipment and operator have long periods of idleness there is a loss of interest on capital.

Transport of machinery
The transport of the machine to the site has not been included. This is because it is best allowed for as a sum in the preliminaries bill. If the plant is owned by the contractor the cost of taking it on to the job would be charged against that job and the cost of removing it against the job it is being transported to. If hired, the cost to and from the job are both charged to that job.

If equipment has a high hire charge and it is not possible to arrange for an unbroken period of use, it may be cheaper to return the plant to the hirer and bring it back to the site at a later date, thus paying double delivery charges rather than pay hire charge for equipment standing idle.

(3) Cost of excavating trenches not exceeding 300 mm wide and not exceeding 2.00 m deep, including removing excavations off site to a tip found by contractor—m³.

Excavations done by backacter owned by contractor and removed by dumpers hired from plant-hire firm.

		£
Cost of machine		*16 000.00*

Standing charges
	Replacement—5 yr	*20%*
	Interest	*10%*
	Repairs	*20%*
		50%

∴ *Total annual cost £8 000.00*

Use 200 days = £40.00 per day or £5.00 per hour

	£
Standing charges per h	*5.00*

Carried forward	*5.00*

$$£$$
Brought forward   *5.00*

Labour operating and attending

	£	
*Driver (1)**	*2.80*	
Labourers (2)	*5.40*	
	——	*8.20*

Fuel

	£	
Diesel, 15 litre at £0.20	*3.00*	
Oil and grease	*0.55*	
	——	*3.55*
		£16.75

Productive rate

0.42 m³ capacity bucket at 40 shovels per h.

40 × 0.42 m³ = 16.80 m³ per h.

$$\text{Cost of excavating per } m^3 = \frac{£16.75}{16.80 \; m^3} = £1.00$$

16.80 m³ excavations required to be removed to a tip 0.5 km distant.

Use two 4 m³ dumpers at a hire rate of £7.00 per h each.

$$£$$
Hire charge, two at £7.00 per h   14.00

Labour operating
Drivers (2) at £2.77 per h   5.54

Fuel
*Diesel 2 × 3 litres = 6 litres
at £0.20   1.20*

£20.74 per h

---

* Drivers of excavators get extra payments depending on the capacity of the machine (National Work Rule 3B sub-para).

Charge for tip—assume free.

$$\therefore Cost\ of\ transport\ per\ m^3 = \frac{£20.74}{16.80\ m^3} = £1.23$$

*Allow for standing time waiting to be
    loaded 10%* = £0.12

                                £1.35

	£
Cost of excavating	1.00
Cost of transport	1.35
	2.35
Profit and oncost 15%	0.35
	£2.70 m³

Cost of excavating and carting £2.70 per m³

It has been assumed that the dumper has a 4 m³ 'struck' capacity and that the amount of the additional material loaded compensates for the 'bulking' of the excavated material.

Dumpers are suitable for short journeys up to ½ km. They are capable of doing two ½ km journeys per h if filled by a mechanical excavator.

(4) Cost of mixing concrete per m³ using a 0.30/0.20(10/7) mixer owned by the contractor.

		£
Cost of mixer		6 000.00
Replacement—8 yr	*12.5%*	
Interest	*10%*	
Repairs	*15%*	
	37.5%	

*∴ Total annual cost = £2 250.00
Use 200 days = £11.25 per day.*

		£
Cost of mixer per day		*11.25*

Labour operating

	£	
Mixer operator	*2.72*	
Cement man	*2.70*	
Aggregate men (*3*)	*8.10*	
per hour	*£13.52*	
per day	*£108.16*	*108.16*

Mixer operator ½ h overtime per day *	*2.04*

Fuel

	£	
Diesel 7 litres at £0.20	*1.40*	
Oil (0.25 litres)	*0.25*	
Waste	*0.10 per day*	*1.75*
	per day £123.20	

Output
Normal day 8 h. Allow 3 min per batch and this gives 160 batches per day.

	160 batches
Less 10%	*16 for breakdowns*
	144
	0.2 m³ per batch
	28.8 m³ per day

$$\text{Cost of mixing per } m^3 = \frac{£123.20}{28.8 \ m^3} = £4.28$$

As this is only a stage in the build-up of a unit rate for concrete the profit and oncost percentage will not be added here but at the end of the complete build-up of price.

* The mixer operator is allowed ½ h overtime for cleaning and preparing the mixer (working rule 3F).

(5) Calculate the cost of mixing concrete per m³ using a 0.30/0.20 mixer hired from a plant-hire firm. (Hire charge £1.65/h.)

Cost of mixer

		£
Hire charge, per day		*13.20*
*Idle time, say 10%**		*1.32*

Labour operating

As Example 4	*108.16*

Fuel

As Example 4	*1.75*

£124.43 per day

Output

As Example 4—28.8 m³ per day.

$$Cost\ of\ mixing\ per\ m^3 = \frac{£124.43}{28.8\ m^3} = £4.32$$

In the previous two examples the cost of mixing concrete has been broken down into a cost per m³. This cost may be added to the unit rate build-up for concrete and therefore gives the cost of concrete including mixing. Alternatively the cost of mixing may be excluded from the unit rate and allowed for as a lump sum in the preliminaries section of the bill of quantities. The period the concrete mixer will be required on site can be determined from the contractor's programme of work in relation to the quantities of concrete involved and the times that they are required. This tends to be the best method, as an accurate lump sum cost of the mixer can be calculated for the contract. In calculating the cost of mixing per m³ however, to a certain extent the non-productive periods of the concrete mixer on site can be charged for by an accurate

* Idle time has been added in this example because for this type of plant the hire cost is not heavy. It is therefore better to keep the mixer on the site over its useful period, even though there are non-productive periods, rather than to increase the transport charges.

assessment of working days per annum in the case of the contractor's own plant, or by a realistic assessment of idle time, in the case of hired plant. This method should give a rate for mixing per m³ that is comparable with the lump sum charge.

(6) Calculate the cost of mixing mortar per m³ using a 0.15/0.10 (5/3½) mixer hired from a plant-hire firm. (Hire charges £6.00 per day.)

Cost of mixer

		£
Hire charge, per day		*6.00*
Idle time, say 10%		*0.60*

Labour operating
    1 operator at £2.70 per h *   21.60 per day*

Fuel		£	
Diesel, 4.5 litres at £0.20		*0.90*	
Oil and waste (say)		*0.20*	
		1.10	
		£29.30 per day	

Output
Normal day, 8 h less ½ h for cleaning and preparation, ie 7½ h running time. Allow 8 batches per h or 60 batches per 7½ h running time.

	60 batches
Less 5%	*3 for breakdowns*
	57
	0.1 m³ per batch
	5.70

Cost of mixing per m³ $= \dfrac{£29.30}{5.70\ m^3} = £5.14$

* This mortar mixer will be serving more than one gang of bricklayers and additional labour will be available in the torm of labourers which are charged for as members of the bricklaying gang.

110

In a similar manner as for mixing concrete the cost of the mixer could be charged for on a lump sum in the preliminaries section on the basis of time the mixer will be required on the site.

Cranes, hoists, etc

The cost of cranes, hoists and similar equipment which perform many tasks, cannot be broken down and allocated to individual unit rates. The estimator must assess the length of time this type of equipment will be required on the site and the total cost is allowed for as a lump sum in the preliminaries section of the bill of quantities.

This lump sum will include for profit and overheads.

Example

Calculate the cost of a rail mounted tower crane (110 tonnes) required on the site for a period of 40 weeks.

	£
Transport to site	*150.00*
Erection costs	
Crane	*600.00*
Track	*225.00*
Hire charges	
Crane: 40 weeks at £20.00 per hour =	
£800.00/week × 40	*32 000.00*
Track: 60 m at £0.30 per m/day =	
£90.00/week × 40	*3 600.00*
Electrical power	
40 weeks at £8.00 per week	*320.00*
Operatives*	
1 crane driver (full-time) and 1 banksman (part-time) allow 40 weeks at £196.00 per week	*7 840.00*
Carried forward	*44 735.00*

* Drivers of cranes get extra wages depending on the capacity of the crane (eg, the driver of a travelling crane gets an extra 12p per hour). A banksman appointed to attend a crane and to be responsible for fastening or slinging loads and generally to direct the crane driver gets an extra 10p per hour.

	£
Brought forward	44 735.00
Temporary base for track including removal and reinstatement of ground, 150 m² at £20.00	3 000.00
Dismantling costs	
Crane	600.00
Track	225.00
	48 560.00
Profit and oncost 15%	7 284.00
Lump sum cost of tower crane for period of contract	£55 844.00

CHAPTER X

PRELIMINARIES

The preliminaries section in a bill of quantities is one of the most important sections that require to be priced. In this section the contractor must allow for all matters affecting his costs that arise out of the conditions of contract, any special requirements of his clients or their professional advisers, and for all temporary works necessary to the carrying out of the contract.

The contractor's tendering strategy will be reflected in the manner in which he prices the preliminaries. All pre-contract planning and tendering decisions require to be made before this section can be fully priced and it is therefore probable that the estimate will be completed prior to this section being priced and the tender figure agreed.

The following is an example of a preliminaries section of a bill of quantities with explanations on the method of pricing:

Page 1
Preliminaries

Item No.	Preliminary particulars		£	p
	Names of Parties			
1	Employer	The Directors, P.Q.R. Co. Ltd.		
2	Architect	Messrs. A.B.C., Chartered Architects		
3	Quantity Surveyor	Messrs. X.Y.Z., Chartered Surveyors		
	Description of Site			
4	The site is situated at the corner of Greenacre Road and Greenacre Terrace, Newtown and is coloured pink on site plan included with these Bills of Quantities.			
5	Access to the Site is by the route indicated on site layout.			
6	The tenderer is recommended to visit the site and inspect trial holes.			

To collection £

Page 2

Item No.		£	
7	**Drawings** The drawings used in the preparation of the Bills of Quantities can be seen, by appointment, or during usual office hours, at the office of the Architect.		
8	**Description of the Works** The work comprises the demolition of part of the existing premises and the building of a new public house of about 25×8 and 5×4 m plan sizes with extension of electrical services and relative external services. The single-storey building has a height of 3.10 m from finished floor level to underside of roof joists. The roof is of flat timber construction covered with bituminous felt roofing. A metal staircase and railing gives access to a roof drying area. The external walls are all cavity walling. The front elevation and return are dressed sandstone facework with horizontal weather-boarded parapet and all other walls are brick. Loadbearing partitions are concrete block or brick and w.c. partitions are breeze block. The floors are concrete with a tile finish in the entrance areas and toilets and a wood boarded finish in the bars. All natural light is by roof light and the cocktail bar has a timber barrel vaulted ceiling. The wall and ceiling finishes are generally plaster. The Contractor, on acceptance of his offer, shall proceed immediately with the preparation of a programme or statement which shall clearly set forth the sequence of all operations and the time limits within which the Contractor proposes that each operation shall be commenced and completed.		
	To collection £		

114

Page 3

Item No.	Clause No.		£	p

The Contractor, in the preparation of this programme shall be held to have co-ordinated the whole of the works embraced in this Contract including the work of nominated sub-contractors, where the necessary information is available to him. On agreement or negotiated amendment of the programme by the Architect, the Contractor shall be responsible for the execution of the works in conformity therewith.

Contract Particulars

9 The works embraced in this Contract are to be carried out in accordance with the Schedule of Conditions of the Standard Form of Building Contract (private edition with quantities) 1963 Edition (July 1977 revision) all of which are held to be incorporated in and form part of the Contract Bills.

10 Execution of the Building Contract shall be deemed to have taken place when letters of offer and acceptance have been exchanged between the parties.

Schedule of Clause Headings

Clause numbers and headings of the Conditions of the Standard Form of Building Contract 1963 Edition (July 1977 revision).

11   1   Contractor's obligations.

12   2   Architect's instructions.

13   3   Contract documents. Clause 3(2) (a) is deleted.

14   4   Statutory obligations, notices, fees and charges (items for rates on temporary buildings and water for the works are provided elsewhere).

15   5   Levels and setting out the works.

16   6   Materials, goods and workmanship to conform to description, testing and inspection.

To collection £

115

Item No.	Clause No.	Page 4	£	p
17	7	Royalties and patent rights.		
18	8	Foreman-in-Charge.	2 300	—
19	9	Access for Architect to the works.		
20	10	Clerk of Works.		
21	11	Variations, provisional and prime cost sums. At Clause 11(4)(c)(i) after "Employers" in line 3 and at 11(4)(c)(ii) after "definition" in line 4 add, "and the Schedule of basic Plant Charges issued by the Royal Institution of Chartered Surveyors".		
22	12	Contract Bills. The Standard Method of Measurement of Building Works 6th Edition shall apply.		
23	13	Contract Sum.		
24	13A	Value Added Tax.		
25	14	Materials and Goods unfixed or off-site.		
26	15	Practical completion and defects liability (see Appendix No 11).		
27	16	Partial completion by employer (see Appendix No 11).		
28	17	Assignment or sub-letting.		
29	18	Injury to persons and property and Employers' Indemnity.		
30	19	Insurance against injury to persons and property (see Appendix No 11).		
31	19A	Excepted risks—Nuclear perils, etc.		
32	20	Insurance of Works against Fire, etc. Clause 20C of the Schedule of Conditions shall apply.		
33	21	Possession, completion and postponement (see Appendix No 11).		
34	22	Damages for non-completion. This clause is not applicable.		
35	23	Extension of time. (Sub-Clause (j) is not to apply).		
36	24	Loss and expense caused by disturbance of regular progress of the work.		
		To collection £	2 300	—

116

m *o.*	*Clause* *No.*		£	p
7	25	Determination by Employer.		
8	26	Determination by Contractor.		
9	27	Nominated Sub-Contractors (see Appendix No 11).		
0	28	Nominated Suppliers.		
1	29	Artists and Tradesmen.		
2	30	Certificates and Payments (see Appendix No 11).		
3	30B	Statutory Tax Deduction Scheme. At the date of tender the Employer was not a 'contractor' for the purposes of the Act.		
4	31	Fluctuations (see Appendix No 11). Clauses 31B, 31C, 31D and 31E shall apply.		
5	32	Outbreak of hostilities.		
6	33	War Damage.		
7	34	Antiquities.		
8	35	Arbitration.		

To collection £

Abstract of Schedule of Conditions

	Clause No.	*Insertion*
Defects Liability Period	15, 16 and 30	12 months
Insurance cover for any one occurrence or series of occurrences arising out of one event	19(1)(a)	£500 000.00
Percentage to cover Professional fees	20(A)	8.5%
Date for Possession	21	To be arranged
Date for Completion	21	6 months from date of possession
Liquidated and Ascertained Damages	22	Not applicable
Period of Delay	26	
(i) by reason of loss or damage caused by any one of the contingencies referred to in Clause 20(A) or Clause 20(B) (if applicable)		3 months

117

(ii) for any other reason		1 month
Prime cost sums for which the Contractor desires to tender	27(g)	To be arranged
Period of Interim Certificates	30(1)	1 month
Retention Percentage	30(3)	10%
Period of Final Measurement and Valuation	30(5)	6 months

Item No.	General Matters	£	p
	Items not covered by the Schedule of Clause Headings:		
49	Plant, tools and vehicles.		
50	Safety, health and welfare of workpeople (including those employed by nominated sub-contractors) employed on the site.		
51	National Insurance and pensions for workpeople.		
52	Holidays for workpeople.		
53	Transport for workpeople.		
54	Safeguarding the works, materials and plant against damage and theft.		
55	Maintenance of public and private roads.		
56	Police regulations.		
57	Noise control.		
58	Overtime will not be refunded unless the Architect gives written instructions to that effect. Where overtime is worked for a specified purpose at the written request of the Architect the Contractor shall be refunded in the 'non-productive' costs of overtime.		
59	Water for the works.	See summary	
60	Temporary arrangements for storing and distributing about the site.		
61	Lighting and power for the works and temporary arrangements for distributing about the site and for lighting to hoardings and the like.		
	To collection £		

	Page 7	£	p

2 The area available for the storage and working of materials is confined, and materials shall be brought onto the site only as required and prepared for incorporation in the buildings as far as possible. Materials shall be deposited only on such areas as allowed by the Architect.

3 The work shall be carried out so as to cause the minimum of interference with the occupants of the premises at which the work is being executed and with any persons using the premises.

Temporary Works

4 Temporary roads, tracks, hardstandings, crossings and the like, including use by nominated sub-contractors.

5 Temporary sheds, offices, messrooms, sanitary accommodation and other temporary buildings for use of the Contractor. *1 150* —

6 Provide temporary dust proof screens about 30 m² where existing walls removed. Provide temporary lockfast doors and barricades to windows in order to ensure maximum security of the building at all times. *90* —

57 Rates on all temporary buildings and temporary works.

58 Temporary fencing, hoardings, fans, planked footways, guard rails, gantries and the like for the proper execution of the work, for the protection of the public and the occupants of the adjoining premises and for meeting the requirements of any local or other authority.

69 General scaffolding for the works.

Protecting, Drying and Cleaning the Works

70 Protecting the works from inclement weather.

	To collection £	*1 240*	—

119

Item No.	*Page 8*	£	p
71	Removing all rubbish and debris from the site and cleaning the works internally and externally; on completion, the works shall be cleaned which shall be deemed to include scrubbing floors, washing pavings, polishing glass inside and out; cleaning sanitary fittings; flushing drains and manholes; cleaning gutters and down pipes; leaving the whole of the new premises clean and ready for occupation.		
	Provisional Sums		
	Provide the following sums for works or costs which cannot be entirely foreseen, defined or detailed:		
72	Temporary equipment, fuel and attendance for drying and controlling the humidity of the works.	*150*	—
73	Builder's work in connection with heating installation.	*200*	—
74	Contingencies.	See summary	
	Provide the following sums for works or costs which cannot be entirely foreseen, defined or detailed and for which the valuation shall be made in accordance with the terms of Clause No 11(4)(c) of the Standard Form:		
75	Prime cost of labour as defined.	*150*	—
	Percentage addition for overheads as defined and profit. *106%*	*159*	—
76	Prime cost of material as defined.	*75*	—
	Percentage addition for overheads as defined and profit *25%*	*18*	*75*
77	Prime cost of plant in accordance with the Schedule of Basic Plant Charges published by RICS.	*35*	—
	Percentage addition for any adjustment of Basic Plant Charges and for overheads as defined and profit *75%*	*26*	*25*
	To collection £	*814*	—

Item No.		Page 9	£	p
		Prime Cost Sums		
78		**Local Authorities or Public Undertakings** Provide the following sums for work to be executed:		
		Water connection.	75	—
		Profit 5%	3	75
		General attendance.	30	—
79		**Nominated Sub-Contractors** Provide the following sums for work to be executed.		
		Heating installation:	1 500	—
		Profit 5%	75	—
		General attendance.	50	—
		Special attendance in connection with oil fired central heating plant and distribution.		
		Unloading and storing materials.	269	—
		Providing power.	20	—
		Protecting the work in this section.	100	—
80		**Nominated Suppliers** Provide the following sums for materials or goods to be supplied:		
		Ironmongery.	200	—
		Profit 2½%	5	—
		To collection £	2 327	75
		Collection **Preliminaries**		
		Amount of page 1		
		Amount of page 2	2 300	—
		Amount of page 3		
		Amount of page 4		
		Amount of page 5		
		Amount of page 6		
		Amount of page 7	1 240	—
		Amount of page 8	814	—
		Amount of page 9	2 327	75
		Amount for Preliminaries carried to Summary £	6 681	75

METHOD OF PRICING PRELIMINARIES ITEMS

Item
No.

1–10 Not priced. These items give general information (the names of the client and his professional advisers, a description of the works and the site and the conditions of contract applicable to the contract). By using the information given in the Description of Works and applying approximate estimating techniques (ie a rate per m²) it is possible to gauge an idea of the value of the works.

11–13 Not priced. These items give general contractual information.

14 Not priced unless there are special instructions for the contractor to include for this in his price.

15 Not generally priced as a preliminary but included as an overhead. The contractor may allow for the setting out of the works in this item if it is more complicated than normal.

16–17 A lump sum to cover the cost of any special testing that is asked for in the bill of quantities, and for the cost of royalties would be allowed for against these items. The lump sum would include for profit and oncosts.

18 Include for the total wages plus additional expenses (eg National Insurance, holidays with pay, etc) of the foreman-in-charge and any other supervisory staff, clerks, time-keepers, etc, who are wholly employed on this contract. The cost of a working foreman may be included in the 'all-in' labour rate as shown previously. The cost of a travelling foreman who is responsible for several contracts may be best allowed for as an overhead charge.
We will assume a foreman-in-charge earning £75.00 per week for the period of the contract (ie 20 weeks). 20 weeks at £75 + $33\frac{1}{3}$ per cent for additional expenses paid by the contractor + 15 per cent for profit and oncosts.

19–25 Not normally priced. These items give general contractual information.

26 Not normally priced. The extent of the defects

liability period will have an effect on the contractor's costs. This may be priced in this item or alternatively the contractor may allow for his total annual costs for maintenance work on contracts as an overhead.

27 The contractor may include here for any additional expense that may occur because of the client's requirement that the work has to be completed in a certain sequence.

28 Not normally priced, as this item gives general contractual information.

29 Not normally priced, as it is usually allowed for under the percentage addition for oncosts.

30 Insurances under 19(1)(a) are normally allowed for in the build-up of the 'all-in' labour rate. Insurance under 19(1)(b) may be priced in the preliminaries but it is more normal for this risk to be covered by a general insurance for all the works to be executed by the contractor and would therefore be covered in the percentage addition for oncosts.

31 Not priced.

32 Not priced in this example as Clause 20C applies. If Clause 20a applies then an insurance specific to the building under consideration should be priced.

33–38 Not priced in this example. If there had been a time limit and liquidated damages for non-completion on time then the contractor would have to consider his commitments fully in order to ascertain if he was capable of fulfilling this condition. In order to complete in time the contractor might require to work overtime and this would affect his overall price. The cost of overtime may be added in the 'all-in' labour rate as described previously or added as a lump sum in the preliminaries. If the contractor wished to tender and knew that he was unable to complete in the time stipulated, he would then have to add to his tender price the cost of liquidated damages for the period he expected to over-run the contract period.

39–40 Not priced here but in the provisional and PC sums section where a percentage for profit should be added to the sum for nominated sub-contractors and suppliers.

Item
No.

41–43 Not priced in this example. If the payments were at different stages of the work or at long intervals then the contractor may charge for interest on the estimated value of the work done for the period it is retained by the client.

44 If the contract is on a firm price basis a sum may be included here to cover anticipated increase in labour and material costs.

45–48 Not priced. These items give general contractual information.

49 Minor plant is usually allowed for when calculating the percentage addition for overheads. Items of plant that can be completely allocated against certain items of work in the bill of quantities may be allowed for in the unit rates for these items. Items of plant that perform general functions and cannot be allocated to specific items in the bill of quantities are normally allowed for as lump sums either here or in the appropriate item in the relative trade section. The lump sum would include for profit and oncosts.

50 Welfare hut, drying accommodation, etc, usually included for under item 65. The cost of wellington boots or protective clothing for workmen may be included here if supplied by the contractor.

51–52 These items are normally allowed for in the build-up of the 'all-in' labour rates.

53 May be allowed for here as a lump sum or included for in the 'all-in' labour rate as shown previously.

54 This item may be priced for in a number of ways:

(i) By allowing for watchmen. The cost would be based on the number of shifts and length of period the watching was necessary plus the cost of accommodation for the watchman.

(ii) By providing a barricade to protect buildings and or materials.

(iii) By allowing a sum of money to cover the cost of any damage or theft from an assessment of past experience from work done in the same area.

55–57 Any anticipated expenses may be allowed for here as a lump sum.

58 Not priced, as this item gives general contractual information.

59 See Summary. This item will be dealt with later.

60 The cost would be allowed for as a lump sum which would be calculated on the basis of the following:
   (i) The number and position of the temporary draw off points required.
   (ii) The cost of cutting tracks and laying temporary piping, including maintenance and removal. The cost of the materials would be less the credit value at removal.
   (iii) The cost of connecting the temporary supply to the nearest permanent water supply pipe.

61 The cost would be allowed for as a lump sum and would include for the following:
   (i) The estimated cost of electricity and electricity authority charge for connection.
   (ii) The cost of all labour and material necessary to provide the temporary supply, including maintenance and removal. The cost of materials would be less the credit value at removal.

62 A sum may be included here for any inconvenience caused and/or double handling required because of the restrictions of the site.

63 A sum may be included here for interference and in phasing the work so as to cause no inconvenience.

64 Reference should be made to the site and drawings to establish the amount of temporary work required. This work would then be costed out, an allowance added for profit and overheads, and the lump sum included in the preliminaries.

65 The requirements for the contract may be calculated as follows:

Cost of storage shed		£
Initial cost		500.00
Replacement—5 years	20%	
Interest	10%	
Repairs	10%	
	40%	

Annual cost = 40% of £500.00 = £200.00

Item
No.   £

    Used 40 weeks per year, therefore cost
    per week is £5.00

Transport to site	15.00
Erection costs	35.00
Used 20 weeks at £5.00	100.00
Dismantling costs	30.00
	180.00
Profit and oncosts 15%	27.00
	£207.00

    The cost of transport off the site would be charged
for as the cost of transport to the site to which the shed
is being transported.

    Contract requirements for sheds, etc:

	£
Shed (storage) 1	207.00
Office (foreman) 1	193.00
Messroom, including drying and dining	
facilities and first-aid	400.00
Sanitary accommodation	350.00
	£1 150.00

66 Allow sum to cover the work as described.
67 The cost of rates would be assessed and allowed for as a
lump sum.
68 Reference should be made to the darwings and the site to
determine the extent of this work.
69 The sum to be included for scaffolding must be carefully
considered as scaffolding requirements vary considerably
from one building to another. The drawings should be
consulted to determine the extent of scaffolding required
and whether putlog or independent scaffolding is necessary. If a large amount of scaffolding is required this
would generally be sub-contracted to a specialist firm and
an estimate of cost would be included under this item.

70 Not generally priced, although a sum for protection may be included if the work involved is expected to be excessive.

71 This may be allowed for in the unit rates or included here as a lump sum of the estimated cost.

Note: The amount of the preliminaries, excluding provisional and prime cost sums, is usually approximately 5 per cent to 10 per cent of the contract figure.

72–74 Provisional sums are for particular works of an unforeseen nature and are amounts of money included by the architect to meet such costs. Contingencies are for unidentifiable work which may be required during the course of the contract. No pricing is necessary by the contractor in respect of these items.

75–77 Provisional sums, for labour, material and plant costs, for works which are to be paid for on the basis of prime cost (ie dayworks) require percentages to be inserted by the contractor which will be added to the cost of each of these items in accordance with the definition for basic costs as contained in the conditions of contract. These percentages are arrived at as follows:

Labour

	Tradesman £	Labourer £
Base rate		
Wage as standard basic rate, joint board supplement, guaranteed minimum bonus: 49 weeks at £70.00 and £60.00	3 430.00	2 940.00
National Insurance 13.5%	463.05	396.90
CITB levy per annum	35.00	6.00
Annual holidays 49 weeks at £8.00	392.00	392.00
Carried forward	4 320.05	3 734.90

127

	£	£
Brought forward	4 320.05	3 734.90

Divide by 1896
hours   Base rate   £2.28/h   £1.97/h

49 weeks × 40
hours   1 960 h

Deduct
Public holidays

8 days × 8 hours   64 h
_____
1 896 h
_____

Allowing two tradesmen to one labourer then the average base rate = £2.18/h
'All-in' labour rate as Note on page 91
Tradesmen £3.00   Labourers £2.70
Contractor's labour costs

	£
'All-in' labour rate allowing 2 tradesmen to 1 labourer	2.90
Full time supervision	0.35
	£3.25
Allowance for nature of work (ie tends to be small quantities and/or disruptive of overall programme of work) 20%	0.65
	£3.90
Overheads and profit 15%	0.59
	£4.49

$$\text{Percentage addition} = \frac{£4.49 - £2.18}{£2.18} \times 100 = 106\%$$

Materials
The percentage is required to cover the cost of

128

PRELIMINARIES

*Item
No.*

overheads and profit plus an allowance for wastage
and additional administration costs.

Percentage addition = 25%

Plant
As this percentage will be applied to the rates
contained in the current Schedule of Basic Plant
Charges then an adjustment is required to bring these
rates up to present day levels. To this figure an
allowance for overheads and profits must be added.
Due to the date when the Schedule of Basic Plant
Charges was last amended it will be necessary to add
about 75 per cent.
Percentage addition = 75%

78   Work, to be carried out by others, for which prime
cost sums are allowed, require to be priced. Under
'profit' an allowance, either as a lump sum or as a
percentage, should be inserted to cover the likely
administrative costs for co-ordinating this work and
also give an element of profit. In this example 5 per
cent has been allowed. 'General attendance' may
include the use of temporary roads, pavings and paths,
standing scaffolding, standing power-operated hoisting plant, the provision of temporary lighting and
water supplies, clearing away rubbish, providing a
suitable space for the sub-contractor's own offices and
for storage for his plant and materials, and for the use
of messrooms, sanitary accommodation and welfare
facilities. In this example £30.00 has been allowed.

79   Prime cost sums for nominated sub-contractors
include a $2\frac{1}{2}$ per cent cash discount (this discount is
dependent on the account being paid within one
month of the statement being received). Where a
nominated sub-contractor is involved then, in
addition to profit and general attendance, there is
frequently special attendance requirements, the
extent of which will vary with the type of work in the
sub-contract.

*Item
No.*

'Profit' and 'general attendance' will be assessed in a similar manner as described for the previous item. The sum for general attendance has been increased due to the longer period of time the sub-contractor's men will be on the site. The 'special attendance', in this example, for the unloading and storing of materials, may involve the supply of a suitable store and labourer's time in unloading and storing materials. The number of deliveries of materials and whether or not men will be available on site, or will require to be sent on site specifically for unloading materials, will influence the amount charged. The costs could be assessed as follows:

	£
Providing, maintaining and removing store	180.00
Labour	
20 h labourer at £2.70	54.00
	£234.00
Profit and overheads 15%	35.00
	£269.00

Under 'providing power' an assessment has to be made on the likely cost involved plus profit. In this example £20.00 has been allowed. 'Protecting the work' would include the cost of protection and/or a sum of money to cover the risks involved. In this example £100.00 has been allowed.

80   Prime cost sums for nominated suppliers include a 5 per cent cash discount (this discount is dependent on the account being paid within one month of the statement being received). The 'profit' item which is to be priced is therefore for charges in addition to the cash discount which, in this case, is included in the £200.00. The sum charged will require to cover administrative costs and profit. In this example it has

been assumed that the accounts will be paid within the stipulated period and an additional charge of $2\frac{1}{2}$ per cent has been made to cover the costs involved.

SUMMARY

The summary appears at the end of the bill of quantities and contains items referred to in the preliminaries. The following is an example of a summary with explanations of method of pricing.

Bill

No.	**Summary**	
1.	Preliminaries	£6 687.75
2.	Excavations and earthwork	£
3.	Etc.	£
A.	Allow for water for the works	Sum
		_____
B.	Allow provisional sum for contingencies which will be under the control of the architect and will be deducted in whole or part if not required without the contractor having any claim for loss of profit	£
		5% _____
	Amount to tender	£ _____

Signed as relative to tender of date

Contractor's Signature ....

Surveyor's address Measured from Plans and Calculated EE

Telephone No. .... (Sgnd) XYZ ....

(Date) .... Chartered Quantity Surveyors

METHOD OF PRICING SUMMARY

Item A   This item would be priced in the summary on the basis of the water authority's charge for supplying water for building works. The water authority's charge is usually on the basis of a percentage of the overall cost of the works, or of the cost of the 'wet' trades portion of the works.

Item B   No costs involved. This figure would be calculated and added into the tender total.

CHAPTER XI

EXCAVATION AND EARTHWORK

Excavation (Mechanical)—outputs

Operation	Plant	Capacity	Output/hour
Reduce level	Tractor crawler	1 m³ shovel	30–40 m³
Basement	Face shovel	0.60 m³	40–50 m³
	Dragline	0.50 m³	25–35 m³
	Backacter	0.50 m³	20–30 m³
Trench	Backacter	0.42 m³	15–20 m³
Pits	Backacter	0.42 m³	10–15 m³

The output depends on the nature of the soil, the area, volume and accessibility of the ground to be excavated.

EXAMPLES

(1)(a) Bring to site and remove from site all plant required for this section of the work—Sum.
(1)(b) Maintain on site all plant required for this section of the work—Sum.

The cost of transporting excavation plant may be in item (1)(a) as a lump sum or alternatively the cost of transporting all plant (including excavation plant) can be allowed for in the preliminaries section under Item No 49 (or other appropriate place).

The cost of maintaining excavation plant may be priced as a lump sum in item (1)(b) or alternatively it can be priced into the rate per hour as shown in Chapter IX (Mechanical Equipment).

(2) Excavate to reduce levels and spread on site average 100 m distant—m³.

Cost of machine (tractor crawler)

		£
Hire charge (per h)		*12.00*
Labour operating	*£*	
Driver (1)	*2.87**	
Banksman (1)	*2.70*	
		5.57

Fuel

Diesel 12 litres at £0.15 = £1.80 per h		
Oil and grease	*£0.30*	
		2.10
		£19.67

Production rate
  30 m³ per hour

$$Cost\ per\ m^3 = \frac{£19.67}{30}$$

	£
	0.66
Profit and oncost 15%	*0.10*
	£0.76 m³

(3) Excavate basement, starting at natural ground level and not exceeding 2.00 m³ deep—m³.

Cost of machine (backacter)

		£
Hire charge (per h)		*12.50*
Labour operating	*£*	
Driver (1)	*2.87*	
Labourer (1)	*2.70*	
		5.57

Fuel

Diesel 10 litres at £0.15 = £1.50		
Oil and grease	*£0.20*	*1.70*
		£19.77

* Drivers of excavators with capacity over ⅜ cu. yd. up to and including ¾ cy. yd. capacity receive an extra payment of 17p per hour. (National Works Rate 3B sub para 7).

Production rate

    20 m³ per hour

	£
$Cost\ per\ m^3 = \dfrac{£19.77}{20}$	*0.99*
Profit and oncost 15%	*0.15*
	£1.14 m³

(4) Surplus excavated material deposited on site in permanent spoil heaps average 50 m from excavations—m³.

Cost of machine (2.30 m³ dumper)

	£
Hire charge per h	*5.00**
Labour operating	
Driver (1) †	*2.77*

Fuel

Diesel 10 litres at £0.15	*= £1.50 per h*	
Oil and grease	*£0.10*	*1.60*
		£9.37

Production rate

    12.50 m³ per hour

	£
$Cost\ per\ m^3 = \dfrac{£9.37}{12.50}$	*0.75*
Profit and oncost 15%	*0.11*
	£0.86 m³

Excavation (Hand)—outputs

Approximate time per m³ for excavating in average soils.

   Excavations thrown and filled into barrows (not exceeding 2.00 m)

* Site use only and therefore excluding road tax and insurance.

† National works rate 3B sub para 6.

Further examples are given in Chapter IX (Mechanical Equipment) and Chapter XXII (Drainage).

	hours
Area	1.0 to 1.75
Trenches	1.75 to 2.25
Pipe tracks	2.5 to 3.25

Allied operations

	hours
Excavate from spoil heaps	0.75 to 1.0
Refill and ram trenches and tracks	1.0 to 1.5
Throw out only (1.5 m stages)	1.25
Wheel 50 m, deposit and return	0.5
Every additional 25 m	0.25
Spread and level	0.4 to 0.75

Note: (i) Excavating for isolated piers, etc, not exceeding 0.5 m² add 25 per cent to trench excavations.

(ii) Excavated materials increase in bulk and this must therefore be allowed for when pricing for carting away.

The outputs for hand excavation to be adjusted by the following percentages depending on the nature of the soil:
Loose soil, −20 per cent; Clay or heavy soil, +50 per cent; Rock, +200 to 300 per cent.

Breaking up existing surfaces

Breaking up macadam surface 150 mm thick, by hand	1.25 man h per m²
Breaking up concrete surface 150 mm thick, by hand	2.5 man h per m²

Earthwork support
Close sheeting

Not exceeding 2.00 m deep	1.25 m² per man per h
Exceeding 2.00 m deep and not exceeding 4.00 m deep	0.75 m² per man per h

Poling boards, waling and struts at 2.0 m centres

Not exceeding 2.00 m deep	2.5 m² per man per h

Hand excavations

An excavator digs a trench to a depth of 1.5 m deep and throws out the excavated soil to the surface for this depth. He may throw the soil to one or both sides and another man may clear away the soil from the edge of the trench. When digging the next 1.5 m in depth the excavator cannot throw the soil to the surface so a platform is provided at the first 1.5 m depth level. He digs and throws the soil onto this platform and another man lifts and throws it to the surface.

For each additional 1.5 m depth the same procedure is carried out.

EXAMPLES

(5) Excavate vegetable soil average 250 mm deep and spread on site average 100 m from excavation—m².

	hour
Excavate and get out	*1.0*
Wheel average 100 m, deposit and return	*1.0*
Spread and level	*0.5*
	2.5
Labourer 2.5 h at £2.70	*£6.75*
Profit and oncost 15%	*1.01*
	£7.76 per m³

£7.76 per m³—therefore 250 mm deep = ¼ of £7.76
= £1.94 per m²

(6) Excavate basements starting from natural ground level and not exceeding 2.00 m deep—m³.

Excavate	*1.5 h*
Labourer 1.5 h at £2.70	*£4.05*
Profit and oncost 15%	*0.61*
	£4.66 per m³

Cost per m³ = £4.66

(7) Excavate trenches to receive foundations, starting from 1 m below natural ground level and not exceeding 2.00 m deep—m³.

136

Excavate	2.0 h

Labourer, 2 h at £2.70	£5.40
Profit and oncost 15%	0.81

£6.21 per m³

(8) Excavate trenches to receive foundations, starting from 1 m below natural ground level, and not exceeding 4.00 m deep—m³.

Excavate	2.25 h
Throw one stage	1.25 h

3.5 h

Labourer, 3.5 h at £2.70	£9.45
Profit and oncost 15%	1.42

£10.87 per m³

(9) Surplus excavated material deposited on site in permanent spoil heaps average 50 m from excavation—m³.

Wheel not exceeding 50 m	0.5 h

Labourer 0.5 h at £2.70	£1.35
Profit and oncost 15%	0.20

£1.55 per m³

(10) Excavated material backfilled around foundations—m³.

Return, fill in and ram	1.25 h

Labourer, 1.25 h at £2.70	£3.38
Profit and oncost 15%	0.51

£3.89 per m³

(11) Earthwork support exceeding 4 m between opposing faces and not exceeding 2 m depth—m².
Timber, £90.00 m³, delivered site.
Calculate on length of excavation 2 m long and 1.5 m deep.

Material

$$m^3$$

Waling 180 mm × 40 mm × 2.0 m   0.0144
Poling 6/180 mm × 40 mm × 1.5 m 0.0648
Struts 2/100 mm × 100 mm × 1.4 m 0.0280
____

		0.1072		£

at £90.00 = 9.65

Nails (say) 0.15
____

£9.80
____

Allow for using 5 times $= \dfrac{£9.80}{5} = £1.96\ per\ 3\ m^2$

£

Materials cost per m²   0.65

Labour
    1 labourer 2.5 m² per h
    £2.82* for 2.5 m²—1 m²   1.13
____

1.78

Profit and oncost 15%   0.27
____

£2.05 † *per m²*
____

(12) Hardcore filling in making up levels over 300 mm thick, deposited and compacted in layers—m³.

---

* A timberman gets £0.12 per hour more than a labourer's wage.

† Earthwork support is measured for all excavations but it is not always required. It is the risk that is being priced more than the labour and materials involved. A rate calculated from experience of about £0.75 per m² would be more realistic depending, of course, on the nature of the soil.

Material

	£	£
Hardcore per m³	*2.00*	
Allow for shrinkage due to		
compaction and waste 25%	*0.50*	
		2.50

Labour

Labourer 1.0 h per m³, 1.0 h at £2.70		*2.70*
		5.20
Profit and oncost 15%		*0.78*
		£5.98 per m³

CHAPTER XII

CONCRETE WORK

Placing concrete per m³

Position	Not reinforced. Hours 1 labourer	Reinforced. Hours 1 labourer
Foundation trenches not exceeding 100 mm thick	2.25	2.75
Ditto exceeding 100 mm thick but not exceeding 150 mm thick	1.75	2.25
Ditto exceeding 150 mm thick but not exceeding 300 mm thick	1.25	1.75
Beds exceeding 150 mm thick but not exceeding 300 mm thick	5.00	6.00
Suspended slabs exceeding 100 mm but not exceeding 150 mm thick	5.75	7.00
Ditto exceeding 150 mm thick but not exceeding 300 mm thick	5.25	6.50
Walls exceeding 100 mm but not exceeding 150 mm thick	6.25	7.75
Ditto exceeding 150 mm thick but not exceeding 300 mm thick	5.25	6.25
Isolated columns not exceeding 0.03 m² sectional area	10.50	11.50
Ditto exceeding 0.03 m² but not exceeding 0.10 m² sectional area	7.75	9.00
Ditto exceeding 0.10 m² but not exceeding 0.25 m² sectional area	6.50	7.75

CONCRETE WORK

Isolated beams not exceeding 0.03 m² sectional area	9.00	10.50
Ditto exceeding 0.03 m² but not exceeding 0.10 m² sectional area	6.50	7.75
Ditto exceeding 0.10 m² but not exceeding 0.25 m² sectional area	5.50	6.50

REINFORCEMENT

Bar reinforcement for reinforced concrete

Mild steel reinforcing bars

Diameter of bar	6 mm	10 mm	12 mm	16 mm	20 mm	25 mm
Weight per kg/m	0.222	0.617	0.888	1.58	2.47	3.86

LABOUR OUTPUTS

Hours 1 steel fixer per tonne

Diameter of bar	20 mm	12 mm	6 mm
Unload and stack on site	3	3	5
Cut to length	3	5	10
Bending to shape	12	17	Bent *in situ*
Fixing	18	20	60
Total per tonne	36	45	75

FORMWORK

Formwork to horizontal soffits of suspended slabs —per m².

Erect and remove (allowing 1 carpenter and 1 labourer one strut per 1.25 m².)   h

Up to 3.50 m high	0.70
3.50 m to 5.0 m high	0.85
5.0 m to 6.5 m high	0.95

Multipliers to be used in conjunction with above
Formwork, per use

One use	1.0
Two uses	0.85

141

Three uses	0.80
Four uses	0.75
Five uses	0.70
Formwork to sloping soffits	
Not exceeding 15° from horizontal	1.15
Exceeding 15° from horizontal	1.25

Formwork to curved surfaces

Small radius	2.15
Large radius	1.50
Struts, two per m²	1.15

Nails

For first use allow 0.5 kg per m² and for each additional reuse allow 0.13 kg per m².

CONCRETE

Composed of cement, sand and aggregate in various portions. Cement weighs 1440 kg/m³ (BSS for ordinary Portland cement). Sand weighs 1600 kg/m³ when dry and approximately 1280 kg/m³ when damp.

Aggregate weighs 1280 kg to 1770 kg/m³, depending on type.

Cost per m³

Rates built up in three sections:
  (i) Cost of materials.
  (ii) Cost of mixing.
  (iii) Cost of placing and compacting.

(a) *Cost of materials*

Cement, £25.00 per tonne; sand, £3.00 per tonne; aggregate, £5.50 per tonne delivered site.

1:2:4 Nominal mix. The actual mix using damp sand would be 1:2½:4.

Materials

Cement	1 part × 1440 kg = 1440 kg	£
	at £25.00 per tonne =	36.00
Sand	2 parts × 1600 kg = 3200 kg	
	at £3.00 per tonne =	9.60
3	*Carried forward*	45.60

142

Brought £
  forward 3 *45.60*
  Aggregate 4 parts × 1650 kg = 6600 kg
                    at £5.50 per tonne = 36.30

      ───
      7 £81.90
Deduct   2.8 shrinkage 40%

      ───
      4.2
      0.2 waste 5%

      ───
      4.0

Cost of materials per m³ = $\frac{£81.90}{4.0}$ = £20.48

(b) *Cost of mixing*
    Hand mixing
    Light aggregate concrete: 1 labourer 4 h per m³.
    Heavy aggregate concrete: 1 labourer 5.25 h per m³.
      In hand mixing a finer mix may require more time, eg a
    1:3:6 mix may take 4.75 h and a 1:2:4 mix may take
    5.25 h. This is not a factor in machine mixing.
    Hand mixing per m³ = 5.25 h at £3.00 = £15.75.

    Machine mixing
    The calculation for machine mixing is shown in Chapter
    IX (Mechanical Equipment).
    Machine mixing per m³ = £4.32.

(c) *Cost of placing*
After the materials are mixed the concrete is transported and
deposited where required. Mixing plant should be positioned
in a place convenient to the point of deposit so the distances
involved are as short as possible. Where long distances are
involved dumpers may be used to transport the concrete. In
contracts where the concrete requires hoisting a mobile or
tower crane and hoppers may be used.

Ready mix concrete
Ready mix concrete is widely used and is very suitable for
congested sites where it is impracticable for the contractor to

set up his own concrete mixing plant or where only small quantities of concrete are required. By calculating the cost of the mixing plant for the period that it is required on the site and the cost of materials it is possible for the contractor to compare the cost of his own mixed concrete with that of ready mix concrete. Where large quantities of concrete are required or where the contractor's own plant would otherwise be lying idle it may be cheaper for the contractor to mix his own concrete.

EXAMPLES

(1) Concrete (1:2:4) in foundations exceeding 100 mm but not exceeding 150 mm thick poured against faces of excavation and laid on earth—m³.

	£
Materials	*20.48*
Mixing	*4.32*
Placing	
1 labourer, 1.25 h at £2.70	*3.38*
	28.18
Profit and oncost 15%	*4.23*
	£32.41 m³

(2) Concrete (1:2:4) in beds exceeding 100 mm but not exceeding 150 mm thick—m³

	£
Materials	*20.48*
Mixing	*4.32*
Placing	
1 labourer, 5.25 h at £2.70	*14.18*
	38.98
Profit and oncost 15%	*5.85*
	£44.83 m³

(3) Reinforced concrete (1:2:4) in isolated beam exceeding 0.10 but not exceeding 0.25 m² sectional area—m³.

		£
Materials		*20.48*
Mixing		*4.32*
Placing		
1 labourer, 6.25 h at £2.70		*16.88*
		41.68
Profit and oncost 15%		*6.25*
		£47.93 m³

REINFORCEMENT

(4) Mild steel bars 12 mm diameter straight and bent reinforcing concrete suspended slabs—tonnes.

Quote: Steel £210.00 per tonne delivered site.

	£
Materials	
Steel, per tonne	*210.00*
Waste and rolling margin 5%	*10.50*
Spacers (say)	*9.00*
Steel tying wire	
13.5 kg per tonne of 12 mm rods at	
£0.50 per kg	*6.75*
Labour	
Steel bender, 45 h at £3.00 (qualified	
benders and fixers are paid the	
craft operators' rate)	*135.00*
	371.25
Profit and oncost 15%	*55.69*
	£426.94 tonne

(5) Steel wire mesh fabric weighing 2.22 kg per m², overlapped 90 mm all joinings—m².

	£
Materials	
Steel fabric 1 m²	*0.60*
Carried forward	*0.60*

145

Brought forward	*£0.60*
Spacers and tying wire (say)	*0.08*
	0.68
Waste and laps 7½%	*0.05*
Labour	
Steel fixer, 0.2 h per m², 0.2 h at £3.00	*0.60*
	1.33
Profit and oncost 15%	*0.20*
	£1.53 m²

FORMWORK

(6) Formwork to horizontal soffits of slabs in 2 No. surfaces—m².

Materials (per 10 m²)	£
9 mm plywood, 10 m² at £4.00	*40.00*
100 mm × 100 mm struts, 0.2 m³ at £95.00	*19.00*
Battens, etc, 0.18 m³ at £95.00	*17.10*
	76.10
Waste 7%	*5.33*
Cost per 10 m²	*£81.43*

Cost per m² = £8.14

Allowing for 1 use

	£
Material cost per m²	*8.14*
Less *reusable value, say 60%*	*4.88*
Actual cost of timber per m²	*3.26*
Nails 0.5 kg at £0.60	*0.30*
Material cost per m² allowing for 1 use only	*£3.56*

Allowing for 5 uses

	£
Material cost, per m²	8.14
Nails:	
1st use, 0.5 kg at £0.60	0.30
4 reuses (allow 0.13 kg per m² for each	
reuse) 4 × 0.13 kg = 0.52 kg at £0.60	0.31
Cost per m² allowing for 5 reuses	£8.75

$$Cost\ per\ m^2 = \frac{£8.75}{5} = £1.75$$

Labour (*allowing for 1 use*)
1 tradesman and 1 labourer = 0.7 h per m².
0.7 h at £5.70 = £3.99 per m².

	£
Materials	3.56
Labour	3.99
	7.55
Profit and oncost 15%	1.13
	£8.68 m²

Labour (*allowing for 5 uses*)
1 tradesman and 1 labourer = 0.49 h per m².
0.49 h at £5.75* = £2.82 per m².

	£
Materials	1.75
Labour	2.82
	4.57
Profit and oncost 15%	0.69
	£5.26 m²

(7) Precast concrete lintels 110 × 230 mm and 2.50 m long, reinforced with one 12 mm diameter mild steel bar—No.

* Allow extra wages for reusing old materials as per working rule.

Mould
No top or bottom required. Calculate on a 1 m lintel.

Materials	£
110 × 32 mm timber 2.5 m at £1.00	*2.50*
Nails, 0.1 kg at £0.60	*0.06*
Bolts, 4 at £0.25	*1.00*
	£3.56
Allow for 25 reuses: £3.56 ÷ 25	*£0.14*

Labour
| *Mould 0.4 h labourer; 0.4 h at £2.70* | *1.08* |
| | *£1.22* |

Reinforcement
Cost of material and fixing £0.43 per kg
(as Example 4) 0.90 kg at £0.43 kg   *£0.39*

Concrete
Cost of material (1:2:4), £20.48 m³ (as
calculated before) 0.03 m³ at £20.48   *0.61*
Labour placing 10 h labourer per m³:
0.03 m³ at £27.00 m³   *0.81*
          £1.42

Hoisting and setting
8 h bricklayer and 16 h labourer per m³.

	£
8 h at £3.00	*24.00*
16 h at £2.70	*43.20*
	£67.20 m³

0.03 m³ at £67.20 = £2.02

148

CONCRETE WORK

	£
Mould	*1.22*
Reinforcement	*0.39*
Concrete	*1.42*
Hoisting and setting	*2.02*
	5.05
Profit and oncost 15%	*0.76*
	£5.81 m

110 × 230 mm precast concrete lintel 2.50 m long —£14.53 No.

CHAPTER XIII

BRICKWORK AND BLOCKWORK

Bricks and mortar per square metre half brick thick
Imperial brick sizes converted to metric:
219 mm × 105 mm × 67 mm ($8\frac{5}{8}''$ × $4\frac{1}{8}''$ × $2\frac{5}{8}''$) and
219 mm × 105 mm × 73 mm ($8\frac{5}{8}''$ × $4\frac{1}{8}''$ × $2\frac{7}{8}''$).

Metric brick sizes:
Clay bricks—215 mm × 102.5 mm × 65 mm: Concrete
bricks—200 mm × 100 mm × 75 mm.

To calculate the number of bricks required per square metre
65 mm bricks with 10 mm beds and joints.

Brick size	215.0	65.0
Bed and joint	10.0	10.0
	225.0	75.0

$$225.0 \times 75.0 = 16\ 875 \text{ mm}^2$$
$$= 60 \text{ bricks per m}^2$$

Mortar requirements per square metre

	Single frog (m^3)	Double frog (m^3)
9.5 mm beds and joints		
67 mm bricks	0.03	0.035
73 mm bricks	0.02	0.025
10 mm beds and joints		
65 mm bricks	0.035	0.04
75 mm bricks	0.025	0.03

Number of facing bricks required per square metre

Bond	Brick size			
	219 × 105 × 67 mm	219 × 105 × 73 mm	215 × 102.5 × 65 mm	200 × 100 × 75 mm
Stretcher	58	53	60	56
English	87	80	90	84

Flemish	77	71	80	75
English				
garden wall	73	66	75	70

Gang size
There is generally 1 labourer to every 2 bricklayers. This may be varied due to circumstances, type or position of work under consideration.

BRICKWORK

Cost of brickwork
The price must include for the following:

(a) The cost, delivered to site, of bricks, cement, lime and sand.
(b) The labour for mixing mortar and building bricks, including labour depositing bricks and mortar at the required positions.
(c) Labour erecting and removing scaffolding as required.
(d) Cost of water.

Items (a) and (b) are included in the unit rate for brickwork.

Items (c) and (d) are included in the preliminaries bill as a lump sum.

Factors affecting the cost of brickwork
(a) The size of the bricks, ie 65 mm, 67 mm, or 75 mm thick. This affects the number to be laid per m^2.
(b) Whether the bricks have frogs, and the position of the frog, affects the amount of mortar required for building.
(c) The mortar mix and the thickness of the beds and joints.
(d) The bond.

COST OF MORTAR

Cement mortar (1:3)
Cement 1 part × 1440 kg = 1440 kg at
£25.00 per tonne = £36.00
Carried forward $\overline{1}$ *36.00*

		£
Brought forward 1		*36.00*

Sand 3 parts × 1600 kg = 4800 kg at
£3.00 per tonne = 14.40

$\overline{}$
4 £50.40

Deduct 0.8 Shrinkage 20%

$\overline{}$
3.2
0.2 Waste 5%

$\overline{}$
3.0

∴ Cost of mortar (materials only) = $\dfrac{£50.40}{3}$ = £16.80 m³

Cement lime mortar (1:2:8)

Cement 1 part × 1440 kg = 1440 kg at
£25.00 per tonne = £36.00

Lime 2 parts × 510 kg = 1020 kg at
£30.00 per tonne = 30.60

Sand 8 parts × 1600 kg = 12 800 kg at
£3.00 per tonne = 38.40

$\overline{}$
11 £105.00

Deduct 2.2 Shrinkage 20%

$\overline{}$
8.8
0.4 Waste 5%

$\overline{}$
8.4

∴ Cost of mortar (materials only) = $\dfrac{£105.00}{8.40}$ = £12.50 m³

Cost of mortar, including mixing = £12.50 + £5.14 = £17.64 m³. (mixing calculated under mechanical equipment chapter)

EXAMPLES

(1) Half brick walls built in cement lime mortar with 10 mm beds and joints—m².

152

Quote: 65 mm common bricks at £30.00 per thousand delivered site.

Material	£
Bricks, 60 at £30.00 per thousand	*1.80*
Waste 5%	*0.09*
Mortar, 0.035 m³ at £17.64 m³	*0.62*

Labour
Bricklayers lay on average between 40 and 100 bricks per h. Assume an output of 60 bricks per h. Using a gang of 4 bricklayers and 2 labourers, a total of 240 bricks would be laid per h.

	£
Bricklayers	
4 at £3.00	*12.00*
Labourers	
2 at £2.70	*5.40*

£17.40 per 240 bricks

$$Cost\ of\ laying\ 60\ bricks = \frac{£17.40}{240} \times 60 = \quad 4.35$$

	6.86
Profit and oncost 15%	*1.03*
	£7.89 m²

Brickwork is generally priced in proportion, eg

Half brick wall	*£7.89 m²*
One brick wall	*£15.78 m²*
One-and-a-half brick wall	*£23.67 m²*

Brickwork in chimney breasts and chimney stacks is generally priced in proportion as the extra labour and height will be compensated by the over-measurement due to the flues not being deducted.

Walls built to radius should be priced more expensive due to increased labour costs.

Walls built to 10 m radius—labour costs increased by 50%.
Walls built to 1.5 m radius—labour increased by 100%.

(2) *Half brick wall, in skin of hollow wall, built in cement lime mortar with 10 mm beds and joints—m².*

Quote: 75 mm concrete bricks at £35.00 per thousand delivered site.

Material	£
Bricks, 56 at £35.00 per 1000	*1.96*
Waste 5%	*0.10*
Mortar 0.025 m³ at £17.64	*0.44*

Labour

Bricklayers laying 60 bricks per h.
Using a gang of 5 bricklayers and 3
labourers a total of 300 bricks will be
laid per h.

Bricklayers	£
5 at £2.22*	*11.10*
Labourers	
3 at £1.94*	*5.82*
	£16.92
Additional*	
indirects 14.5%	*2.45*
	£19.37 for 300 bricks

$$\text{Cost of lay } 56 = \frac{£19.37}{300} \times 56 \qquad 3.62$$

	£6.12
Profit and oncost 15%	£0.92
	£7.04 m²

Cost per m² = £7.04

---

* The labour rates in Example 2 have been based on the example in Chapter VII (Labour Costs) page 90.

(3) Form 50 mm cavity of hollow walls, including three wall ties per square metre—m².

Quote: £5.00 per 100 wall ties

	£
3 No. at £5.00 per 100	*0.15*
Waste 10%	*0.02*
	0.17
Profit and oncost 15%	*0.03*
	£0.20 m²

(4) Close 50 mm cavity of hollow walls with brickwork half brick thick—m.

Material
Allow 5 bricks per metre which includes for waste.

	£
Bricks, 5 at £30.00 per thousand	*0.15*
Mortar (say)	*0.05*

Labour

1 bricklayer with attendant labourer will do 1 m in 0.2 h. Gang as before, 2 bricklayers and 1 labourer	
Bricklayer 0.2 h at £3.00	*0.60*
Labourer 0.1 h at £2.70	*0.27*
	1.07
Profit and oncost 15%	*0.16*
	£1.23 m

Cost per m = £1.23

(5) Extra over common brickwork for facing brickwork, key pointed as the work proceeds—m².
Preambles state all walling to be built in English bond.

Quote: Bricks £35.00 per thousand delivered site.

155

Number of bricks required per m²

Bricks (all stretchers) as calculated before	60
Add for headers, ie double number of bricks	
every second course $= \frac{1}{2}$ *of 60*	30
	90

Material	£
Bricks, 90 at £35.00 per thousand	3.15
Waste 5%	0.16
Mortar 0.04 m³ at £17.64 m³	0.71

Labour

*2 labourers and 4 bricklayers each laying
50 bricks per h, including pointing: 200
bricks per h*

Bricklayers	£
4 at £3.00	12.00
Labourers	
2 at £2.70	5.40

£17.40 per 200 bricks

$$\text{Cost of laying } 90 = \frac{£17.40}{200} \times 90 = \qquad 7.83$$

	11.85
Profit and oncost 15%	1.78
	£13.63

Deduct

Common brickwork allowing for extra	
headers, $1\frac{1}{2} \times £7.89$	11.83
	£1.80 m²

Cost per m² = £1.80

(6) *Extra over common brickwork for facing brickwork, key pointed at a later date—m².*
Preambles state that all walling to be built in Flemish bond.

Quote: Bricks £35.00 per thousand delivered site.

Number of bricks required per m²

Bricks (all stretchers) as calculated before	*60*
Add for headers (alternate headers and	
stretchers) approx. $\frac{1}{3}$ of 60	*20*
	80

Material	*£*
Bricks, 80 at £35.00 per thousand	*2.80*
Waste 5%	*0.14*
Mortar, 0.035 m³ for building	
0.005 m³ for pointing	
0.04 m³ at £17.64 m³	*0.71*

Labour
Laying
2 labourers and 4 bricklayers each laying
55 bricks per h = 220 bricks per h

Bricklayers	*£*
4 at £3.00	*12.00*
Labourers	
2 at £2.70	*5.40*
	£17.40 per 220 bricks

Cost of laying 80 bricks = $\dfrac{£17.40}{220}$ *× 80*   6.33

Pointing
1 bricklayer will point 1.75 m² per h.
1 labourer and 4 bricklayers will point
7 m² per h.

Bricklayers	*£*
4 at £3.00	*12.00*
Labourer	
1 at £2.70	*2.70*
	£14.70 per 7 m²

Carried forward   9.98

		£
Brought forward		*9.98*
$Cost\ per\ m^2 = \dfrac{£14.70}{7} =$		*2.10*
		12.08
Profit and oncost 15%		*1.81*
		£13.89

Deduct
*Common brickwork allowing for extra
    headers, $1\frac{1}{3} \times £7.89$* 10.49

$£3.40\ m^2$

(7) *Half brick walls in skins of hollow walling in multi-colour facing bricks (PC £33.00 per thousand delivered site) in stretcher bond, including flush pointing as the work proceeds—m^2.*

Material   £
    Bricks, 60 at £33.00 per thousand   *1.98*
    Waste 5%   *0.10*
    Mortar, 0.04 m^3 at £17.64 m^3   *0.71*

Labour
    *2 labourers and 4 bricklayers each laying
        50 bricks per h including pointing =
        200 bricks per h.*
    Bricklayers   £
        4 at £3.00   *12.00*
    Labourers
        2 at £2.70   *5.40*

        £17.40 per 200 bricks

$Cost\ of\ laying\ 60 = \dfrac{£17.40}{200} \times 60 =$ 5.22

                                                        8.01
Profit and oncost 15%   1.20

                                                        $£9.21\ m^2$

Cost per m^2 = £9.21

(8) 100 mm concrete block wall built in cement lime mortar—m².

Quote: 440 × 215 × 100 mm blocks—£2.25 per m².

Material   £
    100 mm partition blocks, 1 m²   2.25
    Waste 5%   0.11
    Mortar, 0.015 m³ at £17.64 m³   0.26

Labour
    Bricklayers 0.7 h and labourer 0.35 h
      per m².
    Bricklayers 0.7 h at £3.00   2.10
    Labourer 0.35 h at £2.70   0.95

                                         5.67
    Profit and oncost 15%   0.85

                                       £6.52 m²

(9) Hessian base bituminous sheeting damp proof course bedded in cement mortar on brick walls overlapped 75 mm at all joinings—m².

Material   £
    Damp proof course, 1 m²   0.70
    Cement mortar, 0.01 m³ at £17.64 m³   0.18

                                        0.88
    Waste and laps 7½%   0.07

Labour
    1 h bricklayer and 0.5 h labourer per
      4.25 m².
    £4.35 per 4.25 m²: 1 m²   1.02

                                        1.97
    Profit and oncost 15%   0.30

                                       £2.27 m²

    Cost per m² = £2.27

(10) Hole for large pipe through one brick wall—No.

Material   £

*Bricks and mortar making good round
pipe, included in item for walling*   —

Labour

*0.5 h bricklayer and 0.25 h labourer
per hole*

	£
0.5 h at £3.00	*1.50*
0.25 h at £2.70	*0.68*
	2.18
Profit and oncost 15%	*0.32*
	£2.50 No.

CHAPTER XIV

RUBBLE WALLING AND MASONRY

The following are the weights of various building stones:

	Approx. weight kg/m^3
Sandstones	
Forest of Dean, Gloucestershire	2435
Mansfield, Nottinghamshire	2259
Limestones	
Ancaster, Lincolnshire	2499
Bath, Somerset	2082
Granites	
Rubislaw, Aberdeen	2643
Peterhead, Aberdeenshire	2643

One m^3 of rubble walling requires approximately 0.9 m^3 of stone.

Generally a cement lime mortar is used for building masonry walls.

Approximately 0.2 m^3 of mortar is required per m^3 of rubble walling.

Labour required

Allow 3 labourers for every 2 masons for building rubble walling.

Each mason will build 1 m^3 in 4 hours.

Cement lime mortar (1:2:8)

Calculated in Chapter XIII (Brickwork and Blockwork) page 152.

Cost of mortar = £17.64 per m^3.

161

EXAMPLES
Rubble Walling
(1) 500 mm random rubble wall with natural face built in cement lime mortar—m².

Quote: Limestone £70.00 m³ delivered site.

Material	£
Stone 0.45 m³ at £70.00	*31.50*
Mortar 0.1 m³ at £17.64	*1.76*

Labour		£
2 masons 4 h = 8 h at £3.00	*24.00*	
3 labourers 4 h = 12 h at £2.70	*32.40*	
		56.40
£56.40 for 2 m³		
1 m², 500 mm thick		*14.10*
		47.36
Profit and oncost 15%		*7.10*
		£54.46 m²

Masonry
(2) Granite ashlar 250 mm thick in courses 300 mm high built in cement lime mortar with 10 mm beds and joints, including pare pointing in cement mortar at a later date—m².

Quote: Granite ashlar dressed at quarry £50.00 per m² delivered site.

Material	£
Stone, 1 m²	*50.00*

Mortar
Building 0.06 m³
Pointing 0.006 m³

0.066 m³ at £17.64	*1.16*
Carried forward	*51.16*

	£
Brought forward	*51.16*
Labour	
Building	
2.4 h 1 mason and 1 labourer per m²	
2.4 h at £5.70	*13.68*
Pointing	
0.2 h 2 masons and 1 labourer per m²	
0.2 h at £8.70	*1.74*
	66.58
Profit and oncost 15%	*9.99*
	£76.57 m²

CHAPTER XV

ROOFING

SLATE AND TILE ROOFING

To calculate the number of slates required:

For head-nailed slates: $\text{Gauge} = \dfrac{(\text{length} - 25 \text{ mm}) - \text{lap}}{2}$

For centre-nailed slates: $\text{Gauge} = \dfrac{\text{length} - \text{lap}}{2}$

For plain tiles: $\text{Gauge} = \dfrac{\text{length of tile} - \text{lap}}{2}$

For interlocking tiles: $\text{Gauge} = \text{Length of tile} - \text{lap}$

The gauge is equal to the length of slate visible to the eye. The effective covering area of each slate is therefore the slate width × gauge.

In the case of 406 × 254 mm Welsh slates laid to a 76 mm lap and head-nailed the calculation would be as follows:

$$\text{Gauge} = \frac{(406 - 25 \text{ mm}) - 76 \text{ mm}}{2} = 152.5 \text{ mm}$$

Area covered = 254 × 152.5 mm = 38 735 mm²

$$\text{Number of slates per m}^2 = \frac{1\ 000\ 000}{38\ 735} = 26$$

$$\text{Number of m}^2 \text{ per } 1000 = \frac{1000}{26} = 38.5$$

Allowing for rough edges, etc, say 40 m²

164

Slate nails
38 mm copper or galvanised slate nails weigh 3.18 kg per 1000, ie 320 nails to 1 kg.

Labour outputs
2 slaters and 1 labourer will fix 6 No. 10 m² rolls of felt per h.
2 slaters and 1 labourer will lay 6.75 m² of slating per h (using 406 × 254 mm slates).
1 labourer will unload and stack 1000 slates in 2 h.
1 slater will double hole 1000 slates in 4 h.
1 slater and 1 labourer will sort and double hole 1000 slates in 4 h.
2 tilers and 1 labourer will fix 250 plain tiles double nailed per h; 335 plain tiles single nailed per h; 500 plain tiles hung only per h.

Tile battens
Calculate the length of timber fixed at centres required per m² as follows:
$$\frac{1000 \text{ mm}}{\text{centres}}$$

eg battens at 101.5 mm gauge $= \dfrac{1000}{101.5} = 10$ m.

Allow 1 nail to every 445 mm of timber.

Labour required
2 slaters and 1 labourer will fix 8.5 m² of battens per h at 101.5 mm centres.

EXAMPLES
(1) Cover sarking with underslating felt—m².

Preambles: The underslating felt to be impregnated bituminous felt weighing 13.5 kg per roll of 10 m² to conform to BS747. The felt to be overlapped 75 mm at all joinings and fixed with galvanised nails.

Quote: £4.00 per roll of felt; nails £0.75 per kg; all delivered site. Roll size: 10.00 × 1.00 m.

Actual area covered. Assume three end laps which will reduce the length by 225 mm and one side lap which will

reduce width by 75 mm. Area covered = 9.775 × 0.925
= 9 m².

Materials		£
	Roll of felt	*4.00*
	Nails, 0.15 kg at £0.75*	*0.11*
		4.11
	Waste 2½%	*0.10*

Labour

2 slaters and 1 labourer will fix
6 rolls per h

	£8.70 per 6 rolls: 1 roll	*1.45*
		5.66
	Profit and oncost 15%	*0.85*
		£6.51 per 9 m²

Cost per m² = £0.72

(2) Slate roof 40° pitch with 406 = 254 mm Welsh slates—m².

Preambles: All slates to be machine drilled, double holed and double head-nailed to sarking with 38 mm copper nails and laid to a 76 mm lap.

Quote: Slates £240.00 per 1000; copper nails £0.60 per kg; all delivered site. Number of m² per 1000 slates as previously calculated: 40.

Material		£
	Slate cost per m² = $\dfrac{£240.00}{40}$	*6.00*
	Nails (320 per kg)	
	26 slates per m² = 52 nails	
	0.2 kg at £6.00	*1.20*
		7.20
	Waste 5%	*0.36*
	Carried forward	*7.56*

* Felt is not usually fixed securely as it will be held firmly in position by slate nails when the slates are laid.

£

Brought forward 7.56

Labour

Unloading and stacking:
1 labourer 2 h.
2 h at £2.70 per h = £5.40 per
1000: 26 =   0.14

Holing: 1 slater 4 h.
4 h at £3.00 per h = £12.00
per 1000: 26 =   0.31
Laying: 2 slaters and 1 labourer
6.75 m² per h
6.75 m² cost £8.70: 1 m²   1.29

                                                   9.30
Profit and oncost 15%   1.40

               £10.70 m²

Cost per m² = £10.70

(3) Extra over for double eaves course—m.

Material   £

5 slates per m, including
waste
5 slates at £240.00 per 1000   1.20

Labour

0.75 h tradesman and 0.38 h
labourer per m   £
0.75 h at £3.00   2.25
0.38 h at £2.70   1.03
                              3.28

                              4.48
Profit and oncost 15%   0.67

               £5.15 m

Cost per m = £5.15

(4) Square cutting slating round large openings—m.

Material

		£
Additional waste of about 30 slates per 50 m 30 slates at £240.00 per 1000		7.20

Labour

	£	
20 h slater and 10 h labourer per 50 m		
20 h at £3.00	60.00	
10 h at £2.70	27.00	
		87.00
		94.20
Profit and oncost 15%		14.13
		£108.33 per 50 m

Cost per m = £2.17

(5) Tile roof 45° pitch with 267 × 165 mm concrete plain roofing tiles laid with a 64 mm lap, each tile double nailed with 38 mm copper nails and hung on and including 38 × 19 mm softwood tiling battens—m².

Quote: Tiles £65.00 per 1000; copper nails £6.00 per kg; battens £14.00 per 100 m, all delivered site.

Length of battens required per m² as previously calculated: 10 m.

Number of tiles required:

$$Gauge = \frac{length\ of\ tile - lap}{2} = \frac{267 - 64}{2} = 101.5\ mm$$

Area covered per tile = 165 × 101.5 mm = 16 748 mm².

$$Number\ of\ tiles\ per\ m^2 = \frac{1\ 000\ 000}{16\ 748} = 60$$

$$Number\ of\ m^2\ per\ 1000 = \frac{1000}{60} = 16.67\ m^2$$

Allowing for edge joints not being tight: say 17 m².

Material
Battens

		£	£
	Battens, 10 m at		
	£0.14 per m	*1.40*	
	Nails, 30 nails, 0.1 kg at £6.00	*0.60*	
		2.00	
	Waste 2½%	*0.05*	
			2.05

Tiles

$$\text{Tiles per } m^2 = \frac{£65.00}{17} = \qquad 3.82$$

Nails (320 per kg)
60 tiles per m² = 120 nails
0.4 kg at £6.00 2.40

	6.22	
Waste 5%	*0.31*	
		6.53

Labour
Battens

2 slaters and 1 labourer 8.5 m²
  per h
£8.70 per 8.5 m²: 1 m²   *1.02*

Tiles

Unloading and stacking:
1 labourer 2 h per 1000
2 h at £2.70 = £5.40 per 1000: 60   *0.32*
Laying: 2 slaters and 1 labourer
  250 plain tiles per h
£8.70 per 250 tiles: 60   *2.09*

	12.01
Profit and oncost 15%	*1.80*
	£13.81 m²

Cost per m² = £13.81

(6) Tile roof with 413 × 330 mm concrete interlocking roofing tiles laid with a 76 mm lap, each tile single nailed with 38 mm copper nails and hung on and including 38 × 19 mm softwood tiling battens—m².

Quote: Tiles £175.00 per 1000; copper nails £6.00 per kg; battens £14.00 per 100 m; all delivered site.

Length of battens required per m² $= \dfrac{1000}{337} = 3\ m$

Number of tiles required
Gauge = length of tile − lap* = 413 mm − 76 mm = 337 mm

Area covered per tile = 337 × 292 mm† = 98 404 mm²

Number of tiles per m² $= \dfrac{1\ 000\ 000}{98\ 404} = 10.2$

Number of m² per 1000 $= \dfrac{1000}{10.2} = 98$

Material
Battens

		£	£
Battens 3 m at £0.14		*0.42*	
Nails (allow)		*0.20*	
		0.62	
Waste 2½%		*0.02*	
			0.64

Tiles

		£	£
Tiles per m² $= \dfrac{£175.00}{98}$		*1.79*	
Nails (320 per kg)			
10 nails at £6.00 per kg		*0.19*	
		1.98	
Waste 5%		*0.10*	
			2.08

Labour
Battens

	£
2 slaters and 1 labourer 28 m² per h	
£8.70 per 28 m²: 1 m²	*0.31*
Carried forward	*3.03*

* Interlocking tiles are single lap tiles.
† Width of tile—side lap, ie 330 mm—38 mm = 292 mm.

170

		£
	Brought forward	*3.03*

Tiles

Unloading and stacking:
1 labourer 2.25 h per 1000
2.25 h at £2.70 = £6.08: 10.2   *0.06*
Laying:
2 slaters and 1 labourer 160 tiles per h
£8.70 per 160 tiles: 10.2   *0.55*

          3.64
Profit and oncost 15%   *0.55*

          £4.19 m²

Cost per m² = £4.19

(7) Extra over for black tile ridge bedded and pointed in cement mortar—m.

Quote: 500 mm ridge—£65.00 per 100 m delivered site.

Material   £
     1 length   *0.33*
     Mortar and waste (say)   *0.15*

Labour
     2 slaters and 1 labourer lay
       30 lengths per h
     30 lengths cost £8.70: 1 length   *0.29*

          0.77
Profit and oncost 15%   *0.12*

          £0.89 per 500 mm
                   length
Cost per m = £1.78

CORRUGATED OR TROUGHED SHEET ROOFING

(8) Standard asbestos cement corrugated sheeting to pitched roof fixed with 8 mm galvanised hook bolts and nuts with lead cupped and asbestos washers to steel angle purlins—m².

Quote: Asbestos cement sheets £2.00 per m²; 8 mm galvanised hook bolts and nuts £6.00 per 100; lead cupped washers £1.50 per 100; asbestos washers £0.50 per 100.

Calculated on sheet size of 3.00 × 0.762 and 950 mm purlin spacing.
Allow for 115 mm side laps and 150 mm end laps.
Actual size: 3.00 × 0.762   2.286
Actual area cover: 2.850 × 0.647 1.844

$$0.442 = 20\% \text{ for laps.}$$

Material	£
Sheeting, 1 m²	*2.00*
Allow for laps and waste 25%	*0.50*
Bolts, 3 per m² including waste at	
£8.00 per 100	*0.25*

Labour	
0.3 h tradesman and labourer per m²	
0.3 h at £5.70	*1.71*
	4.46
Profit and oncost 15%	*0.67*
	£5.13 m²

Cost per m² = £5.13

BITUMEN FELT ROOFING

Labour outputs

	1 tradesman and 1 labourer
Nail one layer of felt to timber boarding	*0.07 h/m²*
Lay one layer felt with bitumen up to 10°	*0.10 h/m²*
Lay one layer felt with bitumen exceeding 10°	*0.125 h/m²*
Lay insulation board	*0.04 h/m²*

172

ROOFING

	1 labourer
Lay one coat hot bitumen	*0.20 h/m²*
Spread layer chippings	*0.05 h/m²*

Calculation for laps
Roll size: 10.00 × 1.00 m = 10.00 m²

Actual area covered with 75 mm side and end laps. Assume two laps which will reduce the length by 150 mm and one side lap which will reduce the width by 75 mm.

Actual area covered: 9.85 × 0.925 = 9.11 m²

$$Percentage\ lost\ due\ to\ laps = \frac{10.00-9.11}{9.11} \times 100 = 9.80\%$$

The cost of interest and maintenance on boiler, fuel, and transport of equipment should be allowed for as a lump sum (including profit and overheads) either in the preliminaries section or in the item for protection in the trade section.

(9) Three-layer bitumen felt roofing 75 mm laps, overall bonding between layers and first layer to concrete base, flat over 300 mm wide, and finished with 13 mm granite chippings in hot bitumen—m².

Quote: Felt, £5.00 per roll of 10 m², bitumen; £10.15 kg, granite chips, £6.00 tonne.

Material	£	£
Felt 3 layers at £0.50 per m²	1.50	
Waste and laps 15%	0.23	
		1.73
Bitumen (1½ kg per m² per layer)		
6 kg* at £0.15	0.90	
Granite chips (65 m² per tonne)		
£6.00 per 65 m²: 1 m²	0.09	
	0.99	
Waste 5%	0.05	
		1.04
Carried forward		2.77

* 3 layers for three-layer roofing plus 1 layer for bedding granite chips (ie 4 layers of bitumen).

173

	£
Brought forward	2.77

Labour

Laying three layers of felt and
  bonding:
1 tradesman and 1 labourer 0.3 h
0.3 h at £5.70   1.71
Laying 1 coat of bitumen and
  spreading chips
1 labourer 0.25 h
0.25 h at £2.70   0.68
                                            ─────
                                             5.16
Profit and oncost 15%   0.77
                                            ─────
                                         £5.93 m²
                                            ─────

SHEET METAL ROOFING, FLASHINGS AND GUTTERS

Labour outputs

	Plumber and apprentice per m²	
	2.57 mm thick	2.16 mm thick
Lead flat	2.5 h	2.25 h
Lead gutter	2.75 h	2.5 h

	Plumber and apprentice per m
Lead flashings	0.65 h
Lead wedging	0.1 h
Copper nailing	0.15 h

(10) 2.57 mm sheet lead covering wood flat roof—m².

Quote: 2.57 mm sheet lead £15.00 per m², delivered site.

Material	£
Lead 1 m²	15.00
Waste 2½%	0.38
	─────
Carried forward	15.38

		£
	Brought forward	*15.38*

Labour
    2.5 h 1 plumber and 1 apprentice
    2.5 h at £5.50   *13.75*

                                           29.13
Profit and oncost 15%   *4.37*

                                      £33.50 m²

(11) 2.16 mm sheet lead flashing 150 mm wide and with 150 mm laps, including clips—m.

Quote: 2.16 mm sheet lead £12.50 per m², delivered site.

Material   £
    Lead 1 m 150 mm wide at
        £12.50 per m²   *1.87*
    Allow for laps and clips 5%   *0.09*
    Waste 2½%   *0.05*

Labour
    0.65 h 1 plumber and 1 apprentice
    0.65 h at £5.50   *3.58*

                                       5.59
Profit and oncost 15%   *0.84*

                                       £6.43 m

CHAPTER XVI

WOODWORK

Carcassing—outputs

Classifications (Sectional areas)	Per m³		Per 10 m	
	nails kg	1 carpenter Hours	nails kg	1 carpenter Hours
Floors and flat roofs				
0.005 to 0.008 m²	3.20	14.00	0.20	0.80
0.008 to 0.01 m²	1.60	10.50	0.15	0.90
0.01 to 0.015 m²	1.60	8.75	0.20	1.10
Partitions				
0.004 to 0.005 m²	3.20	15.75	0.15	0.70
Pitched roofs				
0.005 to 0.008 m²	3.20	17.50	0.15	0.80
0.008 to 0.01 m²	3.20	15.50	0.15	0.90
0.01 to 0.015 m²	2.75	14.00	0.12	1.10
Kerbs, beams and the like				
0.003 to 0.005	0.60	9.25	0.05	0.35
0.005 to 0.008	0.60	8.75	0.05	0.25

Per m	Nails kg	1 carpenter Hours
Herringbone strutting	0.20	0.50
Solid strutting	0.10	0.25
Tilting fillets (50 × 40 mm)	0.30	0.15

First fixings—outputs

Per m²	Nails kg	1 joiner Hours
Flooring (T & G)	0.45	0.75
Sarking (butt jointed)	0.30	0.45
Grounds	0.10	0.35

Per m	Nails kg	1 joiner Hours
Eaves and verge boarding	0.30	0.35
Raking cutting boarding	—	0.15
Grounds, plugged, including cutting holes	—	0.14

Second fixings—outputs

Per 100 m	Nails kg	1 joiner Hours
Skirtings	3.50	13.50
Architraves	3.00	15.00
Stops	2.50	10.50

Per m	
Shelving	0.25
Windowboards	1.00
Bearers	0.50

Composite items—outputs

	1 joiner Hours
Making door (based on door 826 × 2040 mm)	
50 mm softwood framed, ledged and braced door	8.00
40 mm softwood four-panel door	10.00
50 mm hardwood single-panel glazed door	10.00
Hanging doors (726 × 2040 mm)	
40 mm softwood door	1.20
40 mm hardwood glazed door	2.00
Making and fitting door frames (based on 726 × 2040 mm door)	
45 × 95 mm to 45 × 145 mm	2.00

Per m²	1 joiner Hours
Making and fitting windows	
Double hung sash and case	10.75
Casement sashes (medium)	8.00

Ironmongery—outputs

	1 joiner *Hours*
Fixing to softwood	
Night latch	1.00
Mortice lock and furniture	2.00
Rim lock and furniture	1.00
Postal knocker	1.50
Casement stays and furniture	0.50

EXAMPLES

Carcassing

(1) 25 × 100 mm Red pine wallplate—m.

Quote: Timber £90.00 per m³ delivered site.

Material		£
	Timber, per m³	*90.00*
	Waste 5%	*4.50*
Labour		
	8.75 h carpenter at £3.00	*26.25*
		120.75
	Profit and oncost 15%	*18.11*
		£138.86

$$Cost\ per\ m = \frac{2.500}{1\ 000.000} \times £138.86 = £0.35$$

(2) 50 × 150 mm White pine in floors—m.

Material		£
	Timber, per m³	*90.00*
	Nails, 3.20 kg at £0.75 per kg	*2.40*
		92.40
	Waste 5%	*4.62*
	Carried forward	*97.02*

	£
Brought forward	*97.02*

Labour
    14 h carpenter at £3.00   *42.00*

	139.02
Profit and oncost 15%	*20.85*

	£159.87

$$\text{Cost per m} = \frac{7.500}{1\,000.000} \times £159.87 = £1.20$$

(3) *50 × 200 mm White pine in flat roofs—m.*

Quote: 50 × 200 mm WP joists—£120.00 per 100 m delivered site.

Material	£
Timber, 10 m at £1.20 per m	*12.00*
Nails, 0.20 kg at £0.75 per kg	*0.15*

	12.15
Waste 5%	*0.61*

Labour
    1.10 h carpenter at £3.00   *3.30*

	16.06
Profit and oncost 15%	*2.41*

	£18.47 per 10m

Cost per m = *£1.85*

(4) *Labour trimming 50 × 150 mm floor joists to chimney breast and hearth 1.5 × 1.2 m overall—No.*

    Allow for 5 joints

Labour	£
1.25 h carpenter at £3.00	*3.75*
Profit and oncost 15%	*0.56*

	£4.31 No.

(5) *50 × 100 mm White pine in pitched roofs—m.*

Material	£
Timber, per m³	*90.00*
Nails, 3.20 kg at £0.75 per kg	*2.40*
	92.40
Waste 5%	*4.62*
Labour	
17.50 h carpenter at £3.00	*52.50*
	149.52
Profit and oncost 15%	*22.43*
	£171.95 per m³

$$Cost\ per\ m = \frac{5.000}{1\ 000.000} \times £171.95 = £0.86$$

(6) *50 × 200 mm White pine in pitched roofs—m.*

Material	£
Timber, per m³	*90.00*
Nails, 2.75 kg at £0.75 per kg	*2.06*
	92.06
Waste 5%	*4.60*
Labour	
14.00 h carpenter at £3.00	*42.00*
	138.66
Profit and oncost 15%	*20.80*
	£159.46 per m³

$$Cost\ per\ m = \frac{10.000}{1\ 000.000} \times £159.46 = £1.59$$

(7) *50 × 100 mm White pine in partitions—m.*

Materials	£
Timber, 10 m at £0.60 per m	*6.00*
Carried forward	*6.00*

	£
Brought forward	6.00
Nails, 0.15 kg at £0.75 per kg	0.11
	6.11
Waste 5%	0.31
Labour	
0.70 h carpenter at £3.00	2.10
	8.52
Profit and oncost 15%	1.28
	£9.80 per 10 m

Cost per m = £0.98

(8) Metal joist hangers handed to builder for building in—No.

Material	£
Hanger 1 No.	0.75
Waste 2½%	0.02
	0.77

Labour
Marking position and attending builder,
30 per h

30 costs £3.00: 1 No.	0.10
	0.87
Profit and oncost 15%	0.13
	£1.00 No.

(9) 25 mm White pine solid strutting between 150 mm deep joists—m.

Material	£
25 × 130 mm timber, 1 m	0.60
Nails, 0.10 kg at £0.60 per kg	0.06
	0.66
Waste 2½%	0.02
Carried forward	0.68

181

	£
Brought forward	0.68

Labour
　　Carpenter 0.25 h at £3.00　　0.75

	1.43
Profit and oncost 15%	0.21

	£1.64 m

First fixings

(10) 25 mm White pine tongued and grooved floor boarding—m².

Material	£
Timber, per m²	3.00
Nails, 0.42 kg at £0.60 per kg	0.25
	3.25
Waste 2½%	0.08

Labour	
Laying and flushing byewood	
Joiner, 0.75 h at £3.00	2.25
	5.58
Profit and oncost 15%	0.84
	£6.42 m²

Cost per m² = £6.42

Note: If this rate had to be worked from a m³ basis then an allowance would require to be made for the tongues. This would, of course, depend on the width of the board, but an average allowance would be 10 per cent to 12 per cent.

(11) 18 mm Flooring grade chipboard, T & G all edges, sealed all round with heavy wax type finish—m².

Material	£
Chipboard, 1 m²	3.00
Nails, 0.10 kg at £0.60 per kg	0.06
	3.06
Waste 5%	0.15
Labour	
Joiner 0.80 h at £3.00	2.40
	5.61
Profit and oncost 15%	0.84
	£6.45 m²

Cost per m² = £6.45

(12) 16 mm White pine boarding to sloping roof—m².

Material	£
Timber, per m³	90.00
Waste 2½%	2.25
	£92.25 per m³

Material cost per m² 16 mm thick =	£
$\dfrac{16}{1\,000} \times £92.25 =$	1.48
Nails, 0.3 kg at £0.75 per kg	0.23
Labour	
Carpenter 0.45 h at £3.00	1.35
	3.06
Profit and oncost 15%	0.46
	£3.52 m²

(13) 230 × 25 mm Red pine fascia plate fixed to ends of rafters—m.

Material	£
Fascia per m	*1.20*
Nails, 0.3 kg at £0.60 per kg	*0.18*
	1.38
Waste 5%	*0.07*
Labour	
Joiner 0.35 h at £3.00	*1.05*
	2.50
Profit and oncost 15%	*0.38*
	£2.88 m

(14) 38 × 25 mm White pine open spaced grounds at 400 mm centres plugged to brick walls—m². Allow for plugs at 600 mm centres.
A 2500 mm length at 400 mm centres covers—2500 × 400 = 1 m².
A length of ground 2.5 m long has 5 plugs.

Material	£
Ground 2500 mm at £0.14 per m	*0.35*
Plugs (say) 5 at £0.03	*0.15*
Nails, 0.1 kg at £0.60 per kg	*0.06*
	0.56
Waste and slips 5%	*0.03*
Labour	
Cutting plug holes, fixing plugs, cutting and fixing grounds.	
Carpenter 0.35 h at £3.00	*1.05*
	1.64
Profit and oncost 15%	*0.25*
	£1.89 m²

Cost per m² = £1.89

(15) 38 × 25 mm White pine grounds plugged to brick walls—m.

Quote: Grounds £15.00 per 100 m.

Material	£
Grounds 100 m	15.00
Nails, 0.80 kg at £0.60 per kg	0.48
	15.48
Waste 2½%	0.39
Plugs (600 mm centres) 168 at £0.05	8.40
	£24.27 per 100 m
Per 1 m	£0.24
Labour	
Joiner, 0.14 h at £3.00	0.42
	0.66
Profit and oncost 15%	0.10
	£0.76 m

Second fixings

(16) 145 × 12 mm Douglas fir skirting once rounded—m.

Quote: 145 × 12 mm skirting, £65.00 per 100 m.

Material	£
Skirting 100 m	65.00
Nails, 3.50 kg at £0.60 per kg	2.10
	67.10
Waste 5%	3.36
Labour	
Joiner, 13.50 h at £3.00	40.50
	110.96
Profit and oncost 15%	16.64
	£127.60 per 100 m

Cost per m = £1.28

(17) 70 × 12 mm Douglas fir architraves, twice rounded—m.

Quote: 70 × 12 mm architraves, £30.00 per 100 m.

Material	£
Architraves 100 m	*30.00*
Nails, 3.00 kg at £0.60 per kg	*1.80*
	31.80
Waste 5%	*1.59*
Labour	
Joiner 15.00 h at £3.00	*45.00*
	78.39
Profit and oncost 15%	*11.76*
	£90.15 per 100 m

Cost per m = £0.90

(18) 230 × 19 mm Douglas fir windowboards—m.

Material	£
Windowboard, 1 m at £0.80	*0.80*
Waste 5%	*0.04*
Nails (say)	*0.05*
Labour	
Joiner, 1.0 h at £3.00	*3.00*
	3.89
Profit and oncost 15%	*0.58*
	£4.47 m

Composite items

(19) 44 mm Afrormosia bound entrance door size 915 × 2058 mm consisting of 95 mm stiles and top rail 195 mm bottom rail, all morticed and tenoned, rebated and prepared for glazing in one pane—No.

Quote: Hardwood £400.00 m³ delivered site.

186

Materials required:
2 × 2.058 = 4.116
            0.915
            _____

        5.031 × 0.100 = 0.503
        _____

        0.915 × 0.200 = 0.183
                    _____

                0.686 × 0.050 = 0.034 m³
Allow 0.036 m³ including waste

Material	£
Hardwood, 0.036 m³ at £400.00 m³	*14.40*
Dressing (allow) £35.00 per m³:	*1.26*
0.036 m³	
	15.66
Allow for wedges, nails, glasspaper and glue 5%	*0.78*

Labour		
Making	*10 h*	
Hanging	*2 h*	
	12	
Joiner, 12 h at £3.00		*36.00*
		52.44
Profit and oncost 15%		*7.87*
		£60.31 No.

Cost per door = £60.31

(20) 40 mm Flush pass door, size 762 × 1980 mm—No.

Material	£
Door, per quotation	*7.50*

Labour	
Fit and hang	
Joiner, 1.2 h at £3.00	*3.60*
Carried forward	*11.10*

	£
Brought forward	*11.10*
Profit and oncost 15%	*1.67*
	£12.77 No.

(21) 3 No. Door frame sets (all same).
120 × 45 mm Red pine jambs plugged to brickwork—m.

Material	£
Frame, 1 m	*0.95*
Plugs (3 per jamb)—allow	*0.08*
	1.03
Waste 5%	*0.05*
Labour	
Joiner, 2 h per door set of 4.80 m	
4.80 m costs £6.00: 1 m	*1.25*
	2.33
Profit and oncost 15%	*0.35*
	£2.68 m

(22) Sash and case window, size 1.00 × 1.75 m comprising sashes 50 mm thick in one pane for glass, 115 × 38 mm stiles and lintel, outer facing 19 mm thick, parting bead 19 × 10 mm, baton rod 25 × 16 mm fixed with brass screws and sockets, twice rebated, twice weathered and twice throated sill 165 × 65 mm, all properly grooved and tongued together complete.

Material   £

50 × 50 mm sashes
$2 \times 2 \times 1.00 \ = \ 4.000$
$2 \times 2 \times 0.875 = \ 3.500$

               7.500 m at £0.50   *3.75*

115 × 38 mm stiles and lintels
$2 \times 1.750 = \ 3.500$
$1 \times 1.000 = \ 1.000$

               4.500 m at £0.65   *2.93*

            Carried forward   *£6.68*

	£
Brought forward	*6.68*
25 × 19 mm outer facings, 4.500 m	
at £0.25	*1.13*
19 × 10 mm parting bead, 4.500 m	
at £0.15	*0.68*
25 × 16 mm baton rod, 4.500 m	
at £0.20	*0.90*
165 × 65 mm sill, 1.000 m	
at £2.35	*2.35*
	11.74
Waste 10%	*1.17*
Allow for wedges, nails, screws,	
glue and sandpaper 5%	*0.59*
	13.50
Labour	
Making, fitting and hanging	
10.75 h joiner per m²	
Joiner, 18.81 h at £3.00	*56.43*
	69.93
Profit and oncost 15%	*10.49*
	£80.42 No.

(23) *Fit and fix Unique spiral sash balances—No.*

Quote: Sash balances £10.00 per pair delivered site.

Material	£
Sash balances, per pair	*10.00*
Labour	
Fitting and hanging per pair	
Joiner, 1.5 h at £3.00	*4.50*
	14.50
Profit and oncost 15%	*2.18*
	£16.68 No.

(24) *White pine staircase 900 mm wide between stringers, and 4.50 m long consisting of 14 steps and one riser, having 230 × 25 mm treads with rounded nosing, 180 × 19 mm riser, tongued into treads and having glued blockings, ends of treads and risers housed into 215 × 32 mm stringers, rounded on one arris, one stringer plugged to brick wall—No.*

Material	£
Treads, 14 No. × 915 mm =	
12.81 m at £1.75	*22.42*
Risers, 15 No. × 915 mm =	
13.73 m at £1.25	*17.16*
Nosing, 1 No. × 915 mm =	
0.92 m at £0.65	*0.60*
Blockings, 14 No. × 3 =	
42 No. at £0.10	*4.20*
Stringers, 2 No. × 4.50 m =	
9 m at £4.00	*36.00*
Plugs, 8 No. at £0.50	*4.00*
	84.38
Waste 10%	*8.44*
Allow for wedges, screws	
and glue 2½%	*2.32*

Labour	
Steps: making, fitting and erecting	
2.25 h per step	
14 No. × 2.25 h = 31.50 h at £3.00	*94.50*
Stringer (plugged)	
Joiner will fix 2.25 m per h	
£3.00 for 2.25 m: 4.50 m	*6.00*
Stringer	
Joiner will fix 3.75 m per h	
£3.00 for 3.75 m: 4.50 m	*3.60*
	199.24
Profit and oncost 15%	*29.89*
	£229.13 No.

(25) *Stock pattern floor unit 600 × 600 × 880 mm fixed to timber framed background—No.*

Material	£
Unit complete	*65.00*
Waste 2%	*1.30*
Labour	
Joiner, 2.0 h at £3.00	*6.00*
	72.30
Profit and oncost 15%	*10.85*
	£83.15 No.

Ironmongery

(26) *102 mm steel hinges to softwood—pair.*

Material	£
Hinge and screws, per pair	*0.75*
Labour	
Included with hanging door	—
	0.75
Profit and oncost 15%	*0.11*
	£0.86 pair

(27) *Fit and fix only 127 mm horizontal mortice lock with furniture and escutcheons to softwood—No.*

Labour	£
Joiner, 2 h at £3.00	*6.00*
Profit and oncost 15%	*0.90*
	£6.90 No.

(28) *Fit and fix only casement stays to softwood—No.*

Labour	£
Joiner, 0.5 h at £3.00	*1.50*
Profit and oncost 15%	*0.23*
	£1.73 No.

CHAPTER XVII

PLUMBING AND MECHANICAL ENGINEERING INSTALLATIONS

Labour outputs

Cast iron

Gutters   *Plumber and apprentice*

100 mm to 150 mm eaves gutters	0.75 h per 1.83 m length
Fittings	0.25 h each

Rainwater pipes

75 mm to 100 mm pipes	0.75 h per 1.83 m length
Fittings generally	0.25 h each

Soil and waste pipes

50 mm to 75 mm pipes	1.25 h per 1.83 m length
100 mm pipes	1.5 h per 1.83 m length
50 mm to 75 mm bends	0.45 h each
100 mm bends	0.75 h each
50 mm to 75 mm branches	0.65 h each
100 mm branches	1.00 h each

PVC

Soil and waste pipes

50 to 75 mm pipes	0.25 h per m
100 m pipe	0.30 h per m
50 to 75 mm bends	0.25 h each
100 mm bend	0.35 h each
100 mm branch	0.40 h each
100 mm bent WC connector	1.00 h each
50 to 75 mm traps	0.50 h each

Copper Tubing

Pipe size	Plumber and apprentice fixing tubing with clips	Plumber and apprentice fixing elbows	fixing tees
10 to 28 mm	0.35 h/m	0.15 h	0.20 h
35 to 42 mm	0.45 h/m	0.20 h	0.25 h
54 mm	0.55 h/m	0.25 h	0.30 h

PVC Tubing

32 mm	0.25 h/m	0.20 h	0.30 h

EXAMPLES

(1)   115 mm Cast iron half-round eaves gutter with red lead and bolted joints in the running length fixed with fascia brackets at 1 m centres—m.

Quote: Gutter £3.75 per 1.83 m length; fascia brackets £0.25 each; red lead £0.35 per kg all delivered site.

Material	£
Gutter—1.83 m length	*3.75*
2 brackets and screws at £0.25	*0.50*
	4.25
Waste 5%	*0.21*
1 joint	
0.1 kg red lead at £0.35	*0.04*
1 bolt and nut	*0.06*

Labour	
0.75 h 1 plumber and 1 apprentice per 1.83 m length	
0.75 h at £5.50	*4.13*
	8.69
Profit and oncost 15%	*1.30*
	£9.99 per 1.83 m length

Cost per m = £5.46

193

(2) Extra over 115 mm cast iron gutter for 115 mm cast iron angle—No.

Quote: Angle £1.50

	£
Material	
Elbow	*1.50*
Waste 2½%	*0.04*
1 joint—as last example	*0.10*
Labour	
0.25 h 1 plumber and 1 apprentice	
0.25 h at £5.50	*1.38*
	3.01
Profit and oncost 15%	*0.45*
	£3.46

	£
Deduct	
450 mm length of 115 mm gutter at	
£5.46 per m	*2.46*
Extra value £1.00 No.	

Extra value for elbow = £1.00

(3) 112 mm PVC half-round gutter, jointed with gutter unions, fixed with and including fascia brackets at 1 m centres—m.

Quote: Gutter £2.20 per 2 m length; gutter unions £0.60 each; fascia brackets £0.25 each; all delivered site.

	£
Material	
Gutter—2 m length	*2.20*
1 gutter union	*0.60*
2 brackets and screws at £0.25	*0.50*
	3.30
Waste 5%	*0.17*
Carried forward	*3.47*

	£
Brought forward	*3.47*

Labour
    0.5 h 1 plumber and 1 apprentice per
      2 m length

0.5 h at £5.50	*2.75*
	6.22
Profit and oncost 15%	*0.93*
	£7.15 per 2 m
	length

Cost per m = £3.58

(4) *100 mm Cast iron coated soil and ventilation pipes with molten lead staved joints in running length with holderbatt fixings to brickwork at 1.83 m centres—m.*

Quote: 100 mm pipes £9.00 per 1.83 m length; lead £1.10 per kg.

Material	£
Pipe—1.83 m length	*9.00*
Holderbatt—1 No.	*0.25*
	9.25
Waste 5%	*0.46*
1 kg lead at £1.10	*1.10*
Yarn (say)	*0.10*

Labour
    1.5 h 1 plumber and 1 apprentice
      per 2 m length   *8.25*

1.5 h at £5.50	
	19.16
Profit and oncost 15%	*2.87*
	£22.03 per 1.83 m
	length

Cost per m = £12.04

(5) *Extra over 100 mm cast iron coated pipe for 100 mm cast iron coated bend—No.*

Quote: 100 mm bend, £3.75.

Material	£
Bend	3.75
Waste 2½%	0.09
1 joint as before	1.20

Labour
0.75 h 1 plumber and 1 apprentice

0.75 h at £5.50	4.13
	9.17
Profit and oncost 15%	1.38
	£10.55

Deduct
300 mm length of 100 mm pipe at

£12.04 per m	3.61
Extra value	£6.94 No.

(6) *90 mm Solid drawn lead soil pipe—m.*

Quote: lead pipe, £9.50 per m delivered site.

Material	£
Pipe 1 m	9.50
Waste 2½%	0.24

Labour
0.6 h 1 plumber and 1 apprentice

0.6 h at £5.50	3.30
	13.04
Profit and oncost 15%	1.96
	£15.00 m

(7) *Solid drawn brass tube ferrule connecting 90 mm lead and cast iron pipe, staved with molten lead including wiped soldered joint—No.*

Material	£
Ferrule	*5.50*
Solder 0.75 kg at £4.00 per kg	*3.00*
	8.50
Waste 1½%	*0.13*
Labour	
1.25 h 1 plumber and 1 apprentice	
1.25 h at £5.50	*6.88*
	£15.51
Profit and oncost 15%	*2.33*
	£17.84 No.

Cost of brass tube ferrule = £17.84

(8) *100 mm PVC soil and ventilation pipework with ring-seal joints in the running length, fixed with supports, measured separately—m.*

Quote: 100 m pipes, including rings, £1.25 per m.

Material	£
Pipe	*1.25*
Waste 2½%	*0.03*
Labour	
0.30 h 1 plumber and 1 apprentice	
0.30 h at £5.50	*1.65*
	2.93
Profit and oncost 15%	*0.44*
	£3.37 m

Cost per m = £3.37

(9) *Extra over 100 mm PVC pipework for 100 mm PVC branch—No.*

Quote: 100 mm branch, £2.00.

	£
Material	
Branch	2.00
Waste 2½%	0.05
Labour	
0.40 h 1 plumber and 1 apprentice	
0.40 h at £5.50	2.20
	4.25
Profit and oncost 15%	0.64
	4.89
Deduct	
350 mm length of 100 mm pipe	
at £3.37 per m	1.18
	£3.71 No.

Extra value for branch = £3.71

(10) *100 mm Standard pipe brackets fixed to background requiring plugging—No.*

Quote: Bracket, £0.20.

	£
Material	
Bracket	0.20
Waste 10%	0.02
Labour	
Included with pipework	—
	0.22
Profit and oncost 15%	0.03
	£0.25 No.

Cost of bracket = £0.25

(11) 22 mm Copper tubing with compression couplings in the running length with clips at 1 m centres screwed to timber—m.

Quote: Tubing £75.00 per 100 m delivered site.

Material	£
Tubing 1 m	*0.75*
1 coupling at £0.65 every 4 m:	
1 m	*0.17*
1 clip with screws	*0.10*
	1.02
Waste 5%	*0.05*
Labour	
0.35 h 1 plumber and 1 apprentice	
0.35 h at £5.50	*1.93*
	3.00
Profit and oncost 15%	*0.45*
	£3.45 m

(12) Extra over 22 mm copper tubing for forming bends—No.

Labour	£
0.15 h 1 plumber and 1 apprentice	
0.15 h at £5.50	*0.83*
Profit and oncost 15%	*0.12*
	£0.95 No.

Cost of form bend = £0.95

(13) Extra over 22 mm copper tubing for 22 mm tee piece—No.

Quote: Compression tee piece £2.00 delivered site.

Material	£
Tee	*2.00*
Waste 1½%	*0.03*
Carried forward	*2.03*

		£
Brought forward		*2.03*
Labour		
0.2 h 1 plumber and 1 apprentice		
0.2 h at 5.50		*1.10*
		3.13
Profit and oncost 15%		*0.47*
		£3.60 No.

Cost of tee piece = £3.60

(14) 635 × 457 mm White glazed lavatory basin with 35 mm diameter chromium plated waste, chromium plated hot and cold water taps and cantilever brackets fitted up complete—No.

Material	£
Basin	*25.00*
2 brackets	*3.50*
Breakages 5%	*1.25*
2 taps at £5.50	*11.00*
35 mm waste plug and chain	*2.25*
Screws (say)	*0.15*
Labour	
2 h 1 plumber and 1 apprentice	
2 h at £5.50	*11.00*
	54.15
Profit and oncost 15%	*8.12*
	£62.27

Cost of basin = £62.27

(15) Galvanised steel cold water cistern with cover 114 litre capacity, holed for two 22 mm and one 28 mm pipes, including 20 mm high pressure ballcock with 150 mm tinned copper ball and lever, and connecting pipes—No.

Material	£
Cistern and cover	*45.00*
Ball valve	*5.00*
	50.00
Waste 1½%	*0.75*

Labour
Fixing cistern 0.75 h
Fix ball valve 0.50 h

1.25 h 1 plumber and 1 apprentice	
1.25 h at £5.50	*6.88*
	57.63
Profit and oncost 15%	*8.64*
	£66.27 No.

Further examples are given in Chapter XXIV (Cost Control).

CHAPTER XVIII

ELECTRICAL INSTALLATIONS

Labour outputs

	Electrician and apprentice
Switchgear—switch unit	0.5 h each
Trunking (mild steel)	3 m per h
Couplings and caps on trunking	0.25 h each
Fixings to concrete	8 No. per h
20 mm steel conduit with clips in chases	3 m per h
1.5 mm² PVC cable in 20 mm conduit	60 m per h
Erect and connect light fittings	1.5 h each
Erect and connect light and power switches	0.25 h each

EXAMPLES

(1) 15 amp 1 way 4 gang surface switch unit fixed with plugs to brickwork—No.

Materials		£	£
	Switch unit	*5.00*	
	Waste 2½%	*0.13*	
		——	*5.13*
Labour			
	0.5 h 1 electrician and		
	1 apprentice		
	0.5 h at £5.50	*2.75*	
	Overheads 20%	*0.55*	
		——	*3.30*
			8.43
	Profit 6%		*0.51*
			——
			£8.94 No.

Cost of switch unit = £8.94

(2) 75 × 50 mm Mild steel lighting trunking, fixed with screws to timber—m.

Materials	£	£
Trunking 1 m	*2.25*	
Waste 5%	*0.11*	
		2.36

Labour		
3 m per h 1 electrician		
and 1 apprentice, 3 m		
cost £5.50		
1 m costs	*1.83*	
Overheads 20%	*0.37*	
		2.20
		4.56
Profit 6%		*0.27*
		£4.83 per m

Cost per m = £4.83

(3) Extra over 75 × 50 mm mild steel trunking for tee junction—No.

Materials	£	£
Tee piece	*2.20*	
Waste 5%	*0.11*	
		2.31

Labour		
0.25 h 1 electrician and		
1 apprentice		
0.25 h at £5.50	*1.38*	
Overheads 20%	*0.28*	
		1.66
		3.97
Profit 6%		*0.24*
		£4.21 No.

Cost per tee junction = £4.21

(4) *20 mm Heavy gauge steel screwed conduit fixed to brick in chases—m.*

Quote: 20 mm conduit—£0.85 per m.

Materials		£	£
Conduit 1 m		0.85	
Waste 5%		0.04	
Allow for fittings			
50%	£0.45		
Waste 2½%	0.01		
		0.46	
			1.35

Labour			
3 m per h 1 electrician			
and 1 apprentice			
3 m cost £5.50			
1 m costs		1.83	
Overheads 20%		0.37	
			2.20
			3.55
Profit 6%			0.21
			£3.76 per m

Cost per m = £3.76

(5) *1.5 mm² PVC cables in conduit—m.*

Quote: PVC cable £400 per 100 m.

Materials	£	£
PVC cable 1 m	0.04	
Waste 5%	0.001	
		0.041

Labour		
60 m per h 1 electrician		
and 1 apprentice		
60 m cost £5.50		
1 m costs	0.09	

Carried forward	0.09	0.041

		£
		£
Brought forward	£0.09	0.041
Overheads 20%	0.02	
	——	0.11
		0.151
Profit 6%		0.009
		£0.16 per m

Cost per m = £0.16

(6) *Circuit wiring in 1.5 mm² PVC insulated and sheathed twin core and earth cable to lighting points —No. (For local authority housing.)*

Materials	£	£
PVC cable 6 m at 0.20	1.20	
Waste 5%	0.06	
	——	1.26
Labour		
15 m per hour 1 electrician and 1 apprentice		
15 m cost £5.50		
6 m cost	2.20	
Overheads 20%	0.44	
	——	2.64
		3.90
Profit 6%		0.23
		£4.13 No.

Cost per lighting point = £4.13

(7) *Circuit wiring in 1.5 mm² PVC insulated and sheathed twin core and earth cable to 1 way switch points, grouped as required—No. (For local authority housing.)*

Materials	£	£
PVC cable 5 m at £0.20	1.00	
Waste 5%	0.05	
	——	1.05
Carried forward		1.05

		£
	Brought forward	*1.05*

Labour
> *15 m per hour 1 electrician*
>   *and 1 apprentice*
> *15 m cost £5.50*

5 m cost	*1.83*	
Overheads 20%	*0.37*	
	——	*2.20*
		3.25
Profit 6%		*0.20*
		£3.45 No.

Cost per switch point = £3.45

(8) Erect and connect up pendant light fitting—No.

Labour

1.5 h 1 electrician and 1 apprentice	*£*
1.5 h at £5.50	*8.25*
Profit and oncost 26%	*2.15*
	£10.40 No.

Cost of connecting up fitting = £10.40

206

CHAPTER XIX

FLOOR, WALL AND CEILING FINISHINGS

Labour outputs for plastering per square metre

	1 plasterer and 1 labourer
Render and set on walls (2 coats)	0.5 h
Render and set on ceiling (2 coats)	0.55 h
Render, float and set on walls (3 coats)	0.6 h
Render, float and set on ceilings (3 coats)	0.65 h

In widths not exceeding 300 mm wide add 100%.

Labour outputs for lath per square metre

	1 plasterer and 1 labourer
Fixing plaster lath	0.25 h
Fixing metal lath	0.35 h

First coat plaster to walls

	mm	mm	mm	mm	mm
Finished thickness when applied	9.5	13	16	19	25
Thickness for estimating purposes:					
Brick walls	16	19	25.4	31.8	38
Rubble walls	19	22.2	28.6	44.5	47.5

The second coat for three-coat work should be considered as 9.5 mm and the setting coat as 6.4 mm.

General labour outputs

	1 plasterer and 1 labourer
Forming arrises	3.75 m per h

Fixing metal corner beads	13.5 m per h
Run plaster cornices exceeding 150 mm but not exceeding 300 mm girth	1.0 m per h

Hardwall plaster

Proportions by volume	1:1	1:2	1:3
Equivalent proportions by weight	1:1½	1:3	1:4½

Applications of hardwall plaster on brick walls (Two coats 12.7 mm thick)

Floating coat composed of 1 part browning to 3 parts sand by volume scratched to provide good key.

Finishing coat should be finish, applied neat or with an admixture of well-slaked putty lime not exceeding 25 per cent volume trowelled to a smooth surface.

Covering capacity of 1 tonne of hardwall plaster

	Proportion by volume	*m²*
On clay brick walls	Floating coat (1:3)	225–250
On concrete brick or blocks	Floating coat (1:2)	175–190
On plasterboards	Floating coat (1:1½)	205
On metal laths	Floating coat (1:2) and Rendering (1:1½)	115
Neat finish	—	370–410

Covering capacity of 1 tonne of Carlite plaster

		m²
Carlite browning		
On brick walls, clinker partitions, etc	11 mm	140
Carlite metal lathing		
On expanded metal	8 mm	60–65
On wood wool slabs	11 mm	120
Carlite bonding coat		
On concrete and on plaster boards	8 mm	150
Carlite finish	1.6 mm	410–490

EXAMPLES
(1) Expanded metal lath fixed to framing at 300 mm centres—m².

208

Materials	£
Metal lath 1 m²	*0.70*
Waste and laps 10%	*0.07*
Staples (say)	*0.12*

Labour
    0.35 h 1 plasterer and 1 labourer

0.35 h at £5.70	*2.00*
	2.89
Profit and oncost 15%	*0.43*
	£3.32 m²

(2) Gypsum lath and two coats hardwall plaster on ceiling—m².

Quote: Lath £0.50 per m²; nails £15.50 per 25 kg; browning £37.00 per tonne; finish £35.00 per tonne; sand £4.50 per tonne, all delivered site.

Preambles: The floating coat to be composed of 1 part browning to 1½ parts sand by volume, scratched to receive finishing coat. The finishing coat to be finish hardwall plaster applied neat with a mixture of not more than 25 per cent of volume of putty lime.

Lath

Material	£	£
Lath 1 m²	*0.50*	
Nails (21 per m²), 0.06 kg at		
£0.62	*0.04*	
	0.54	
Waste 2½%	*0.01*	

Labour
    0.25 h 1 plasterer and 1 labourer

0.25 h at £5.70	*1.43*	
		1.98

Carried forward   *1.98*

209

		£
	Brought forward	1.98

Plaster
  Material £
  Floating coat
    1 tonne browning at
      £37.00   37.00
    2 tonne sand at £4.50   9.00
                              ─────────
                               £46.00
                              ─────────

  This covers 205 m²
  $1 \ m^2 = \dfrac{£46.00}{205}$ 0.22

  Finish £
    1 tonne finish at £35.00   35.00
    0.25 tonne hydrated lime
      at £30.00   7.50
                              ─────────
                               £42.50

  *Labour preparing 0.25
    tonne hydrated lime to
    putty lime 2 h
    labourer*
  2 h at £2.70   5.40
                              ─────────
                               £47.90
                              ─────────

  This covers 450 m²
                                              0.11
  $1 \ m^2 = \dfrac{£47.90}{450}$ ─────
                                               0.33

  Waste 5%   0.02
Labour
  *0.55 h 1 plasterer and
    1 labourer*
  0.55 h at £5.70   3.14
                              ─────────   3.49
                                         ─────────
                                           5.47
                                         ─────────

Carried forward 5.47

	£
Brought forward	*5.47*
Profit and oncost 15%	*0.82*
	£6.29 m²

Cost per m² = £6.29

(3) 2 coats Carlite plaster on brick walls—m².

Quote: Browning £55.00 per tonne; finish £45.00 per tonne, delivered site.

Material	£
Floating coat	
1 tonne browning at £55.00	
This covers 140 m², 1 m² $= \dfrac{£55.00}{140}$	*0.39*
Finishing coat	
1 tonne finish at £45.00	
This covers 450 m², 1 m² $= \dfrac{£45.00}{450}$	*0.10*
	0.49
Waste 5%	*0.02*

Labour	
0.5 h 1 plasterer and 1 labourer per m²	
0.5 h at £5.70	*2.85*
	3.36
Profit and oncost 15%	*0.50*
	£3.86 m²

Cost per m² = £3.86

(4) 2 coats Carlite plaster on brick walls not exceeding 300 mm wide—m².

Material	£
As Example 3	*0.51*
Carried forward	*0.51*

211

		£
Brought forward		*0.51*

Labour
    As Example 3 plus 100%
    £2.85 + 100%   *5.70*

                                6.21
    Profit and oncost 15%   *0.93*

                          £7.14 m²

Cost per m² = £7.14

(5) Metal corner beads on external corners—m.

Material   £
    Metal beads 1 m   *0.20*
    Waste 5%   *0.01*

Labour
    1 plasterer and 1 labourer 13.5 m per h
    £5.70 per 13.5 m: 1 m   *0.42*

                                  0.63
    Profit and oncost 15%   *0.09*

                          £0.72 m

Cost per m = £0.72

(6) 2 coats cement plaster (1:3) on brick walls finished 19 mm thick—m².

Quote: Cement £27.00 per tonne; sand £4.50 per tonne, delivered site.

Material   £
    Cement 1 part × 1440 = 1440 kg at
                  £27.00 per tonne   *38.88*
    Sand 3 parts × 1600 kg = 4800 at
                 .   *£4.50 per tonne*   *21.60*

                  ––
Carried   4
forward   *60.48*

		£
Brought forward	*4.0*	*60.48*
Deduct	*0.8 Shrinkage 20%*	

3.2

0.2 Waste 5%

3.0 £

$Cost\,of\,materials\,only = \dfrac{£60.48}{3} =$ *20.16 m³*

Labour (mixing)
*Hand mixing, 1 labourer 8 h per m³ * *
8 h at £2.70   21.60

£41.76 m³

Required thickness is 19 mm, therefore estimating thickness is 31.8 mm.

Cost of 1 m² 31.8 mm thick at £
£41.76 m³.
£41.76 × 0.0318 = £1.33 m²   1.33

Labour
0.45 h 1 plasterer and 1 labourer per m²
0.45 h at £5.70   2.57

3.90
Profit and oncost 15%   0.59

£4.49 m²

(7) Plaster cornice girth 230 mm—m.
Material

£
Plaster stucco (say)   0.70

Carried forward   0.70

* If machine mixing is required then the cost would be calculated in a similar manner to that of mortar as shown in Chapter IX (Mechanical Equipment) page 110.

		£
Brought forward		*0.70*

Labour

1.0 h 1 plasterer and 1 labourer
    per m
    1 h at £5.70   *5.70*

                                                  6.40
Profit and oncost 15%   *0.96*

                                              £7.36 m

(8) Form arrises on plaster—m.

Labour   *£*

3.75 m 1 plasterer and 1 labourer per h
3.75 m cost £5.70: 1 m   *1.52*
Profit and oncost 15%   *0.23*

                                              £1.75 m

Cost per m = £1.75

(9) 3 coats roughcast on brick walls—m².

Quote: Granite chippings £6.50 per tonne, delivered site.

Material   *£*

First 2 coats as 2 coats cement plaster   *1.33*
Dashing coat (per 100 m²)   *£*
1.2 tonnes chippings at £6.50   *7.80*
0.25 tonnes cement at £27.00   *6.75*

                      100 m² £14.55

1 m²   *0.15*

                                              1.48
Waste 5%   *0.07*

Labour
2 rendering coats 0.45 h

Carried forward   *0.45*   **£1.55**

		£
Brought forward	*0.45*	*£1.55*
Dashing coat	*0.30 h*	

0.75 h 1 plasterer and
1 labourer

0.75 h at £5.70	*4.28*

	5.83
Profit and oncost 15%	*0.87*

	£6.70 m²

Cost per m² = £6.70

(10) Cement and sand screed (1:3) 25 mm thick finished smooth on top—m².

Material   £

Cost of material per m³ £41.76, as
 calculated for cement plaster.
Require 25 mm, therefore estimate for
 32 mm thick.
Cost of 1 m² 32 mm thick at £41.76
 per m³
£41.76 × 0.032 =   *1.34*

Labour
0.5 h 1 plasterer and 1 labourer per m²
0.5 h at £5.70   *2.85*

	4.19
Profit and oncost 15%	*0.63*

	£4.82 m²

Cost per m² = £4.82

(11) Granolithic 25 mm thick finished smooth on top—m².

Preambles: 2 parts cement; 1 part sand; 3 parts granite chips.

215

Material   £

  Cement 2 parts × 1440 kg = 2880 kg
      at £27.00 tonne   77.76

  Sand   1 part 1600 kg = 1600 kg
      at £4.50 tonne   7.20

  Granite 3 parts × 1760 kg = 5280 kg
      at £6.50 tonne   34.32

   $\dfrac{}{6}$ £119.28

Deduct   1.5 Shrinkage 25%

    $\dfrac{}{4.5}$
    0.2 Waste 5%

    $\dfrac{}{4.3}$

   Cost of material only $= \dfrac{£119.28}{4.3} =$ £27.74 m^3

Labour (mixing)
  Hand mixing, 1 labourer 8 h per m^3
  8 h at £2.70   21.60

           £59.34 m^3

Require 25 mm, therefore estimate for 32 mm thick.
  Cost of 1 m^2 32 mm thick at £59.34
   per m^3 £
  £59.34 × 0.032   1.90

Labour
  0.5 h 1 plasterer and 1 labourer per m^2
  0.5 h at £5.70   2.85

           4.75
  Profit and oncost 15%   0.71

           £5.46 m^2

(12) Quarry tiles to BS 1286 type A size 100 × 100 × 15 mm bedded and jointed in cement mortar (1:3)—m^2.

Quote: Tiles, £75.00 per 1000 delivered site.

Material
>   *Cost of cement mortar per m³ £41.76 as calculated for cement plaster.*
>   *Require 12 mm, therefore estimate for 19 mm thick.*
>   *Cost of 1 m² 19 mm thick at £41.76 per m³*

		£
£41.76 × 0.019		*0.79*
Grouting and pointing: ⅓ *of £0.79*		*0.26*
		1.05
Tiles, 100 at £75.00 per 1000	*£7.50*	
Waste 2½%	*0.19*	
		7.69

Labour
>   *1.75 h 1 tiler and 1 labourer*
>   *1.75 h at £5.70* — 9.98

	18.72
Profit and oncost 15%	*2.81*
	£21.53 m²

Cost per m² = £21.53

For an example on dry dash finish to walls see Chapter XXV (Analogous (Pro-rata) Rates) page 217.

CHAPTER XX

GLAZING

GLAZING (WITHOUT BEADS) PER SQUARE METRE

	Putty kg	1 Glazier h
Steel sashes		
Not exceeding 0.10 m²	1.20	1.85
Over 0.10 but not exceeding 0.50 m²	1.00	1.40
Over 0.50 but not exceeding 1.0 m²	0.50	0.65
Over 1.0 m²	0.45	0.45
Wood sashes		
Not exceeding 0.10 m²	1.00	1.65
Over 0.10 but not exceeding 0.50 m²	0.75	1.20
Over 0.50 but not exceeding 1.0 m²	0.45	0.55
Over 1.0 m²	0.40	0.35

EXAMPLE

(1) 3 mm sheet glass (OQ) in steel sashes with putty in panes exceeding 0.10 but not exceeding 0.50 m²—m².

Quote: 3 mm glass £6.00 per m²; putty £0.30 kg; all delivered site.

Material		£
	Glass 1 m²	*6.00*
	Putty 1 kg at £0.30	*0.30*
		6.30
	Waste 15%	*0.95*
Labour		
	1.40 h tradesman	
	1.40 h at £3.00	*4.20*
	Carried forward	*£11.45*

Brought forward   *£11.45*

Profit and oncost 15%   *1.72*

                                               £13.17 m²

Cost per m² = £13.17

(2) *Insulight hermetically sealed units comprising two panes 4 mm clear float (GG) with one 13 mm air space in pane size 2000 × 1000 mm in wood sash with glazing compound—No.*

Material		£
	Double glazing unit	
	(2.00 × 1.00 m)	*40.00*
	Glazing compound 0.75 kg	
	at £0.30	*0.23*
		40.23
	Waste 10%	*4.02*
Labour		
	Tradesman 1.25 h per m²	
	2.50 h at £3.00	*7.50*
		51.75
	Profit and oncost 15%	*7.76*
		£59.51 No.

CHAPTER XXI

PAINTING AND DECORATING

MATERIALS

50 kg of water paint will cover about 250 m² of plasterwork in one coat.

2 kg of putty will stop about 100 m² of woodwork.

0.7 litre of varnish will knott about 100 m² of woodwork.

10.5 litres primer will cover about 100 m² depending on nature of base.

8 litres undercoating will cover 100 m².

5.5 litres finishing coat will cover about 100 m².

LABOUR OUTPUTS	Painter per h
Prepare and first coat water paint on walls	6.75 m²
Second coat water paint on walls	10 m²

Painting 100 m² of woodwork

	Painter per 100 m²
Knotting	4 h
Stopping	5 h
Priming, including preparing	18 h
Undercoating (per coat)	15 h
Finishing coat	14 h

Wallpaper

A piece of English wallpaper measures 10 m long by 533 mm broad. Waste allowances are generally 10 per cent on plain paper and 15 per cent on pattern paper.

	Paperhanger per piece
Sandpaper and size walls	0.5 h
Wallpaper	1.25 h

	Paperhanger per m²
Strip existing paper and prepare walls	0.25 h

EXAMPLES
*(1) One coat primer, one coat undercoating and one coat gloss
paint on plaster walls—m².*

Material		£	£
Primer, 10.5 litre at £1.30		*13.65*	
Undercoat, 8 litre at £1.15		*9.20*	
Finish, 5.5 litre at £1.20		*6.60*	
		———	*29.45*
Waste 5%			*1.47*

Labour
Priming, including

Preparing	*16 h*
Undercoating	*15 h*
Finishing coat	*14 h*
	——
	45 h

45 h at £3.00	*135.00*	
Allow for waste of		
brushes, 5%	*6.75*	
	———	*141.75*
		———
		172.67
Profit and oncost 15%		*25.90*
		——— *per*
		£198.57 100 m²

Cost per m² = £1.99

(2) Two coats emulsion paint on plaster walls—m².

Material		£	£
First coat	*8 litres*		
Second coat	*5.5 litres*		
	——		
	13.5 litres		
13.5 litres at £1.00 litre		*13.50*	
Waste 5%		*0.68*	
		———	*14.18*

Labour
*Prepare and 2 coats
paint 30 h per 100 m²*

30 h at £3.00	*90.00*	
Carried forward	*90.00*	*14.18*

	£	£
Brought forward	90.00	14.18
Allow for waste of brushes 5%	4.50	
	——	94.50
		108.68
Profit and oncost 15%		16.30
		—— per
		£124.98 100 m²

Cost per m² = £1.25

(3) One coat primer, one coat undercoating and one coat gloss paint on new woodwork—m².

Material	£	£
Knotting 0.7 litre at £1.95	1.37	
Putty 2.0 kg at £0.30	0.60	
Primer 10.5 litre at £1.30	13.65	
Undercoating 8 litres at £1.15	9.20	
Finish 5.5 litres at £1.20	6.60	
	——	
	31.42	
Waste 5%	1.57	
	——	32.99

Labour		
Knotting	4 h	
Stopping	5 h	
Priming	18 h	
Undercoating	15 h	
Finishing	14 h	
	—	
	56	
	—	

56 h at £3.00	168.00	
Allow for waste of brushes 5%	8.40	
	——	176.40

Carried forward 209.39

	£
Brought forward	209.39
Profit and oncost 15%	31.41
	£240.80 per
	100 m²

Cost per m² = £2.41

(4) Ditto, ditto on skirtings and the like, exceeding 150 mm but not exceeding 300 mm girth—m.

Material		£
As Example 3		32.99
Labour		
As Example 3	176.40	
Add 20% for cutting*	35.28	
		211.68
		244.67
Profit and oncost 15%		36.70
		£281.37 per
		100 m²

Cost per m² = £2.81
Cost per m (average 250 mm wide) = £0.70

(5) Pattern wallpaper on plastered walls including preparing and sizing (prime cost value £5.00 per piece)—m².

Material	£	£
Wallpaper 1 piece	5.00	
Paste 0.25 kg at £0.75	0.19	
Sandpaper and size (say)	0.25	
	5.44	
Waste 15%	0.82	
		6.26
Carried forward		6.26

* The 20 per cent for cutting has been added to allow for extra cutting to line at change of colour on two edges.

223

$£$

Carried forward   6.26

Labour
       Preparing   *0.5 h*
       Papering   *1.25 h*
          ‾‾‾‾
          1.75 h
          ‾‾‾‾

1.75 h at £3.00	5.25	
Allow for waste of brushes	0.26	5.51
5%	‾‾‾‾	‾‾‾‾
		11.77
Profit and oncost 15%		1.77
		‾‾‾‾
		£13.54 per piece
		‾‾‾‾

1 piece of wallpaper $= 5\frac{1}{3} m^2$
Cost per $m^2 = £2.54$

CHAPTER XXII

DRAINAGE

LABOUR OUTPUTS
For labour outputs for excavations see Chapter XI (Excavation and Earthwork).

AVERAGE WIDTHS OF DRAIN TRACK EXCAVATIONS

	Diameter of pipe	
	Up to *200 mm*	*200 mm* *to* *300 mm*
Track up to 2.00 m deep	530 mm	700 mm
Track 2.00 to 4.00 m deep	610 mm	760 mm

FIRECLAY PIPES (CEMENT MORTAR JOINTS)

Diameter	Yarn kg	Cement mortar litre³	Labour laying and jointing 1 m pipes man h
100 mm	0.014	0.5	0.15
150 mm	0.027	0.9	0.16
225 mm	0.053	1.7	0.25

Cast iron

	Plumber and apprentice
100 mm spigot and socket pipes	0.95 h/m
100 mm bend	1.45 h
Lead: 2 kg per joint	

Unplasticised PVC

110 mm single socket pipe	1.20 h per 6 m length
110 mm branch	0.40 h each

PIPE TRENCH EXCAVATIONS

Mechanical excavation

		£
Cost of machine (backacter)		
Hire charge per h		10.50

Labour operating:

	£	
Driver (1)	2.80	
Labourer (1)	2.70	
		5.50

Fuel:

Diesel 12 litres at £0.15	1.80 per h	
Oil and grease	0.20	
		2.00
		£17.00

Production rate
12 m³ per hour
   (excavate and backfill)

	£
Cost per m³ $= \dfrac{£17.00}{12}$	1.42
Profit and oncost 15%	0.21
	£1.63 m³

Hand excavation

Rates for excavations are calculated in a similar manner to that for foundation trench excavations.

The m³ rate for pipe trench excavations is first calculated, eg pipe trench excavations not exceeding 1.50 m deep—m³.

Excavate, get out and remove surplus	2.5 h
Refill and ram	1.0 h
	3.5 h

226

	£
Labourer 3.5 h at £3.00	10.50
Profit and oncost 15%	1.58
	£12.08 m³

EXAMPLES

(1) Excavate trench not exceeding 2.00 m deep and average 1.25 m deep for drain pipes not exceeding 200 mm diameter, earthwork support, filling, compacting, grading bottom to correct falls, and remove surplus material—m.

Volume of excavations per m of trench:
1.0 × 0.53 × 1.25 deep = 0.663 m³

Hand excavation:
Volume of trench excavation per m = 0.663 m³ at
£12.08
= £8.00 m

Machine excavation:
Volume of trench excavation per m = 0.663 m³ at £1.63
= £1.08 m*

Allow for earthwork support if considered necessary. The cost may be calculated in a similar manner to that shown in Chapter XI, Example 11.

Drain pipes and fittings
Cost of mortar (1 part cement:
1 part sand) for jointing:

Material	£
Cement 1 part × 1440 kg = 1440 kg	
at £25.00 per tonne	36.00
Sand 1 part × 1600 kg = 1600 kg	
at £3.00 per tonne	4.80
	£40.80

Carried forward   2

---

* If filling by hand is required then this rate would be increased considerably.

Brought forward   2.0

Deduct   *0.4 Shrinkage 20%*

                      $\overline{1.6}$

                      0.3 Waste 20%

                      $\overline{1.3}$

                                            £

Cost of material per $m^3 = \dfrac{£40.80}{1.3}$ *31.38*

Labour

    Hand mixing, 1 labourer 8 h per m^3

    8 h at £2.70   *21.60*

                                    £52.98 m^3

Cost per $litre^3 = £0.053$

(2) *100 mm Salt glazed fireclay drain pipes laid and jointed with (1:1) cement mortar—m.*

Quote: 100 mm pipes delivered site—list price £0.75 + 5% + 55% per m.

Material

	£	£
100 mm drain pipe	0.75	
Plus 5%	0.04	
	$\overline{0.79}$	
Plus 55%	0.43	
	———	1.22
1 joint		
0.5 $litre^3$ *mortar at £0.053*	0.027	
0.014 kg yarn at £2.70	0.038	
	———	0.07
		$\overline{1.29}$
Waste 5%		0.06
		$\overline{}$
Carried forward		1.35

	£
Brought forward	*1.35*

Labour
> *1 drainlayer and 1*
> > *labourer 0.15 h per m*
>
> *0.15 h at £5.47**

	0.82
	———
Profit and oncost 15%	2.17
	0.33
	———
	£2.50 m

(3) *Extra over 100 mm fireclay drain pipes for 100 mm salt glazed fireclay branch piece—No.*

Quote: 100 mm branch piece delivered site—list price £1.50 + 5% + 55%.

Material	£	£
100 mm branch piece	1.50	
Plus 5%	0.08	
	———	
	1.58	
Plus 55%	0.87	
	———	2.45
2 joints		
2 joints as Example 2–		
2 at £0.07		0.14
		———
		2.59
Waste 2½%		0.06
Labour		
Laying and making 2 joints, 1 drain-		
layer and 1 labourer 0.23 h at £5.47		1.26
		———
		3.91
Profit and oncost 15%		0.59
		———
		4.50
Carried forward		4.50

* Pipelayers and jointers for pipes not exceeding 300 mm diameter get an extra 7p per hour.

	£
Brought forward	*4.50*

Deduct

    Branch displaces 1.0 m of 100 mm
       drain pipe at £2.50 per m   *2.50*

    Extra value for branch piece   **£2.00 No.**

(4) 100 mm Cast iron spun drain pipes, BS 1211 Class 'B', with molten lead staved joints in the running length, laid in trenches—m.

Quote: 100 mm pipes £13.50 per 1.83 m length; lead £1.10 per kg.

Material   £

	£
Pipe 1.83 m length	*12.00*
Waste 5%	*0.60*
1 kg lead at £1.10	*1.10*
Yarn (say)	*0.10*

Labour

    1 plumber and 1 apprentice 0.95 h
      per m
    1.74 h at £5.50   *9.57*

	23.37
Profit and oncost 15%	*3.51*

                per 1.83 m length   **£26.88**

Cost per m = £14.69

See also similar example in Chapter XXIV (Cost Control).

(5) 110 mm Unplasticised PVC drain pipe with joints in the running length, laid in trenches—m.

DRAINAGE

Material	£
Pipe 6 m length	*10.00*
Ring seal joint	*0.50*
	10.50
Waste 5%	*0.53*

Labour
1 plumber and 1 apprentice
1.20 h per 6 m length
1.20 h at £5.50 .... *6.60*

Profit and oncost 15% *17.63*
.... *2.64*

per 6 m length **£20.27**

Cost per m = £3.38

CHAPTER XXIII

ALTERATIONS

Work in alterations is generally priced higher than similar items of new work, depending of course on the nature and extent of the particular contract under consideration. Generally the reasons for higher prices are as follows:

(1) Work in alterations tends to be in small areas and/or quantities which do not allow the same level of outputs as can be achieved with larger areas or quantities. The time taken is also longer because of the need to cut and/or bond between new and existing work. The wastage of materials may also be greater due to the extra cutting and the small quantities involved.

(2) The work may be non-continuous (ie not in large areas but scattered throughout the existing building). This can affect the labour costs for preparing and transporting materials.

(3) The work may be less accessible than would be the case in a building in course of erection. Travel distances for men and materials may be greater because of the nature of the existing building.

(4) Care must be taken not to damage existing surfaces as they will require to be made good at the end of the contract. An allowance to cover the likely cost of this work could be included in the preliminaries section.

(5) The use of plant may be restricted and/or it may be less effectively utilised than in new works.

(6) Work in alterations does not lend itself to the introduction of incentive bonus schemes and this may influence both the length of the contract and the labour costs.

(7) Profit being based on turnover, then the longer the time taken to complete the works then the longer the time required to earn this money. Profit levels should

therefore be increased to take account of extra time requirements in relation to contract values (eg if a contract for £50 000 normally takes six months if new work, and nine months if mainly alterations work, and the percentage charged for profit is the same in both instances, then both contracts will earn the same amount of profit. In order to compensate for the extra time involved and to make a more efficient use of the firm's resources the profit requirements should be adjusted).

Value Added Tax

Work in new construction, alterations or demolitions are at present zero rated for VAT. Repair and maintenance works, however, are positive rated at 15 per cent. In alterations work it can be difficult to distinguish between the two categories. For example, if in the course of alteration works an existing casement window is removed and replaced by a new casement window, then this work is subject to VAT. If however the window opening is enlarged and a new casement window fitted, then the work would be zero rated.

CHAPTER XXIV

COST CONTROL

The estimate, as mentioned previously, should reflect the work done on the site. The following example is intended to illustrate how a small contractor can use his estimate to achieve a better understanding and control of labour hours and material costs on the site. Instead of building up unit rates on a vertical basis as has been previously described, pre-lined estimating sheets are used and the estimate is computed on a horizontal basis. In this example the waste allowances on materials has been taken as 5 per cent on pipework and $2\frac{1}{2}$ per cent on fittings; overheads and profit have been assumed to be 25 per cent on both material and labour costs. By reference to Figure 4 it can be seen that by this method it is possible to obtain the total number of labour hours, the cost of materials and the charge for overheads and profit which have been included in this estimate for the work. These totals may be further broken down into sub-totals (such as cast iron, PVC, copper, etc) which may facilitate better control.

A comparison of the actual number of hours taken, for example, with the number of hours included in the estimate, should assist both with the control of the work on site within the estimate price and enable outputs to be adjusted for estimating future works where they are seen to be unrealistic in relation to the figures actually achieved on the site. Good estimating based on the feedback of information from the site and cost control within the amounts allowed in the estimate are prerequisites for a healthy business.

EXAMPLE

Extract from the Plumbing and Mechanical Engineering Installations Section of a bill of quantities.

Item No.	Sanitary installation	Rate			£	p
	Supplied appliances Fit and fix; connect pipes					
1	Water closet set with low level cistern	No.	2			
	Drainage pipework, cast iron pipes and fittings BS416 Type A sockets					
	Pipes; coated inside and outside by manu- facturer; caulked lead joints in the running length, laid in trenches					
2	100 mm internal diameter	M	10			
3	Extra; branches	No.	2			
	Soil and vent pipework; PVC pipes and fittings, Osma or equal and approved					
	Pipes; ring-seal joints in the running length, fixed in supports measured separately					
4	100 mm internal diameter	M	20			
	Standard pipe brackets					
5	For 100 mm pipes; fixed to backgrounds requiring plugging	No.	12			
	Heated water installation					
	Service pipework; copper pipes BS 2871 Table X; fittings BS 864					

(continues page 238)

Item No.	BQ quantity	Material	Material costs (incl waste) £	£	Labour outputs h	Total labour hours h	h
1	2 No.	—	—	—	2.50	5.00	
							5.00
2	10M	Cast iron drain at £6.30	63.00		0.95	9.50	
		5 kg lead at £1.10	5.50				
		Yarn	1.00				
				69.50			
3	2 No.	CI branch at £8.20	16.40		1.75	3.50	
		4 kg lead at £1.10	4.40				
		Yarn	0.40				
				21.20			

Deduct
500 mm/branch = 1.00 CI drain at £15.22

Item No.	BQ quantity	Material	Material costs (incl waste) £	£	Labour outputs h	Total labour hours h	h
4	20M	PVC pipe at £2.31	46.20	46.20	0.30	6.00	
5	12 No.	Brackets at £0.26	3.12	3.12	Incl		19.00
6	20M	28 mm copper at £1.58	31.60		0.35	7.00	
		20 No. clips at £0.06	1.20				
		6 No. unions at £0.50	3.00				
				35.80			
7	5 No.	28 mm made bend	—	—	0.20	1.00	
8	2 No.	28 mm elbows at £0.77	1.54		0.15	0.30	
9	3 No.	28 m Tee at £3.59	10.77		0.20	0.60	
10	20M	Insulation at £1.68	33.60		0.60	12.00	
				45.91	—		20.90
		Totals		£221.73		hours	44.90

Labour costs £	£	Material and labour costs £	£	Overheads + profit £	Total £	BQ rate £	BQ quantity	BQ totals £
7.50		27.50	27.50	6.88	34.38	17.19	2 No.	34.38
—	27.50	——						
2.25		121.75		30.44	152.19	15.22	10M	152.20
9.25		40.45		10.11	50.56		2 No.	35.34
					−15.22			
					35.34	17.67		
33.00		79.20		19.80	99.00	4.95	20M	99.00
—		3.12		0.78	3.90	0.33	12 No.	3.96
——	104.50	——	244.52					
38.50		74.30		18.58	92.88	4.64	20M	92.80
5.50		5.50		1.38	6.88	1.38	5 No.	6.90
1.65		3.19		0.80	3.99	2.00	2 No.	4.00
3.30		14.07		3.52	17.59	5.86	3 No.	17.58
66.00		99.60		24.90	124.50	6.23	20M	124.60
——	114.95	——	196.66					
	£246.95	£468.68	£117.19					£570.76

Figure 4. Build-up of unit rates for bill of quantities.

237

(continued from page 235)

Item No.				£
		Pipes; couplers in the running length; fixed to timber background with copper saddles		
	6	28 mm	M 20	
		Extra; made bends	No. 5	
		Extra; elbow	No. 2	
		Extra; brass tee	No. 3	
		Insulation 19 mm Rigid fibreglass insulation, canvas finish complete with metal fixing bands To 28 mm pipes	M 20	

MATERIAL QUOTATION—ALL D/d

100 mm Cast iron drain pipe	£6.00 m
100 mm Cast iron drain branch	£8.00 each
100 mm PVC pipe	£2.20 m
100 mm Standard bracket	£0.25 each
28 mm Copper tube	£1.50 m
28 mm Brass elbow	£0.75 each
28 mm Brass tee	£3.50 each
19 mm Fibreglass insulation for 28 mm pipes	£1.60 m

CHAPTER XXV

ANALOGOUS (PRO RATA) RATES

Variations in building contracts in which a bill of quantities forms the basis of the contract are normally priced at bill rates or at rates in strict accordance with bill rates. The determination of these pro rata or analogous rates requires a prior knowledge of building estimating procedure. In order to build up a new rate for agreement between the surveyor and the contractor, the surveyor must first analyse the appropriate bill rates in order to establish either the labour involved or the percentage to add for profit and oncost. The new rate is built up on the same basis as the analysed rate and the new labour rate or the calculated percentage for profit and oncost added.

EXAMPLES
Method 1—To establish the labour content.

(1) Three coats roughcast (19 mm thick), with white Portland cement and having granite chip wet dash finishing coat—bill rate £6.44 m².

Breakdown of bill rate:

	£
Bill rate	*6.44*
Deduct	
*Profit and oncost (15%), ie 13%**	*0.84*
Net cost of labour and material	*£5.60*

---

* In order to calculate the true amount of profit and oncost that has been added to the cost of materials and labour to arrive at a unit rate, the deduction from the unit rate would be one-ninth, 13 per cent and one-sixth for profit and oncosts of 12½ per cent, 15 per cent and 20 per cent respectively.

Material   £

White cement 1 part × 1440 = 1440 kg
       1440 kg at £55.00 per tonne   79.20
  Shiver sand 3 parts × 1600 = 4800 kg
       4800 kg at £5.00 per tonne   24.00

               4 parts   £103.20

Deduct

          0.80 shrinkage 20%

          3.20
          0.20 waste 5%

          3.00

$$Cost\ of\ materials\ per\ m^3 = \frac{£103.20}{3.00} = £34.40\ m^3$$

Machine mixing: cost of mixer allowed for in preliminaries as a lump sum for all works in this section of the bill of quantities.
Required thickness is 19 mm, therefore estimating thickness is 31.80 mm.

                                 £

Cost of 1 m² 31.80 mm thick at £34.40 m³
= £34.40 × 0.0318   1.09

Dashing coat (per 100 m²)	£
1.20 tonnes granite chippings	
at £6.50	7.80
0.25 tonnes white cement	
at £55.00	13.75
	21.55
Waste 5%	1.08
100 m²	£22.63

                      Carried forward   1.09

	£
Brought forward	*1.09*
1 m²	*0.23*
Cost of materials:	**£1.32**

Cost of labour $= £5.60 - £1.32 = £4.28\, m²$

Three coats roughcast (19 mm thick) having white spar chip dry dash finishing coat—m².
Build up of unit rate:

Materials £

*First two coats as breakdown
    of cost*   1.09
*Dashing coat
1 tonne of spar chippings
    covers 75 m²*   £

$1\, m²\, costs\, \dfrac{£15.00}{75\, m²}$ 0.20

Waste 10%   0.02   0.22

Labour

*1 plasterer and 1 labourer per m²
Wet dash—0.75 h
Dry dash—0.65 h
Labour in ratio of 75 : 65*

$Labour\, cost = \dfrac{65}{75} \times £4.28$ 3.71

    5.02

Profit and oncost 15%   0.75

    £5.77 m²

Pro rata rate $= £5.77\, m²$

Method 2—To establish the percentage allowed for profit and oncosts.

(2) Break down of bill rate.
Extra over common brickwork for facings PC £40.00 per 1000

241

in English bond, key pointed as the work proceeds—bill rate £2.94 m².

Number of bricks, 10 mm beds and joints (calculated	
before)	*60*
Add for headers, ie double number of bricks each	
alternate course, ½ of 60	*30*
	90

Material		£
Bricks, 90 at £40.00 per 1000		*3.60*
Waste 5%		*0.18*
Mortar, 0.04 m³ at £17.64		*0.71*

Labour

4 bricklayers and 2 labourers laying
    50 bricks per h—200 bricks per h

Bricklayers	*£*	
4 at £3.00	*12.00*	
Labourers		
2 at £2.70	*5.40*	
Per 200 bricks	*£17.40*	

$$\text{Cost of laying } 90 = \frac{£17.40}{200} \times 90 = \qquad 7.83$$

	12.32
Percentage required for profit and	
oncost 20%	*2.46*
	14.78

Deduct

Common brickwork allowing for headers:	
1½ × £7.89* per m²	*11.84*
Bill rate	*£2.94 m²*

---

* From Example 1 in Chapter XIII (Brickwork and Blockwork).

Build up of pro rata rate.
Extra over common brickwork for facings PC £50.00 per thousand in Flemish bond, key pointed as the work proceeds—m².

Number of bricks, 10 mm beds and joints (calculated before)	*60*
Add for headers, alternate headers and stretchers, $\frac{1}{3}$ of 60	*20*
	80

	£
Material	
Bricks, 80 at £50.00 per 1000	*4.00*
Waste 5%	*0.20*
Mortar, 0.04 m³ at £17.64	*0.71*
Labour	
4 bricklayers and 2 labourers laying 45*	
bricks per h = 180 bricks per day	
As before £17.40 per 180 bricks	
Cost of laying $80 = \dfrac{£17.40}{180} \times 80 =$	*7.73*
	12.64
Add profit and oncost as calculated	
previously 20%	*2.53*
	15.17
Deduct	
Common brickwork allowing for headers:	
$1\frac{1}{3} \times £7.89$ per m²	*10.49*
	£4.68 m²

Pro rata *rate* = *£4.68 per m²*

---

* Bricklayers more accustomed to laying bricks in English bond, therefore the output will be greater than that for Flemish bond.

INDEX